Springer Series in Optical Sciences

Volume 192

Springer Series in Optical Sciences

The Springer Series in Optical Sciences, under the leadership of Editor-in-Chief William T. Rhodes, Georgia Institute of Technology, USA, provides an expanding selection of research monographs in all major areas of optics: lasers and quantum optics, ultrafast phenomena, optical spectroscopy techniques, optoelectronics, quantum information, information optics, applied laser technology, industrial applications, and other topics of contemporary interest.
With this broad coverage of topics, the series is of use to all research scientists and engineers who need up-to-date reference books.
The editors encourage prospective authors to correspond with them in advance of submitting a manuscript. Submission of manuscripts should be made to the Editor-in-Chief or one of the Editors. See also www.springer.com/series/624

More information about this series at http://www.springer.com/series/624

Victor V. Apollonov

High-Power Optics

Lasers and Applications

Springer

Victor V. Apollonov
Department of High-Power Lasers
General Physics Institute of the Russian
Academy of Sciences
Moscow
Russia

ISSN 0342-4111 ISSN 1556-1534 (electronic)
ISBN 978-3-319-10752-3 ISBN 978-3-319-10753-0 (eBook)
DOI 10.1007/978-3-319-10753-0

Library of Congress Control Number: 2014950677

Springer Cham Heidelberg New York Dordrecht London

Printed on acid-free paper

Springer is part of Springer Science+Business Media (www.springer.com)

Preface

Based on the theory we developed in the early 1970s, a broad range of phenomena is considered for an optical surface of a solid body that is exposed to radiation arbitrarily varying in time and producing temperature gradients, thermoelastic stresses and thermal deformations on the surface layer. The examination is based on the relations between the quantities characterising the thermal stress state in any nonstationary regimes of energy input into a solid, which are similar to Duhamel's integral formula from the theory of heat conduction. A peculiar feature of the analysis of the thermal stress state in this case consists in the fact that this relation comprises time as a parameter, which in turn is a consequence of incoherence of the quasi-stationary problem of thermoelasticity. This phenomenon is particularly important for the optics of high-power, high-pulse repetition rate lasers, which are being actively developed. In the review we have recently published in Laser Physics, the thermal stress state of a solid is analysed. In this state, time is treated as an independent variable used in differentiation. Such an approach greatly reduces the possibility of the method. The review published contains data on the use of capillary porous structures made of various materials with different degrees of surface development. Moreover, such structures can be efficiently employed to increase the heat exchange at a temperature below the boiling point of the coolant. In the present review we discuss the dependences of the limiting laser intensities on the duration of a pulse or a pulse train, corresponding to the three stages of the state of the reflecting surface and leading to unacceptable elastic deformations of the surface, to the plastic yield of the material accompanied by the formation of residual stresses and to the melting of the surface layer. We also analyse the problem of heat exchange in the surface layer with a liquid metal coolant pumped through it. The theoretical estimates are compared with the experimental data. We discuss the issues related to the technology of fabrication of power optics elements based on materials with a porous structure, of lightweight highly stable large optics based on highly porous materials, multi-layer honeycomb structures and silicon carbide, as well as problems of application of physical and technical fundamentals of power optics in modern high-end technology.

These cooling techniques can be applied to other technologies. Consider the integration of a large number of laser diodes into one- and two-dimensional array structures radiating high-power laser radiation either incoherently or coherently with respect to each other. One difficulty of such arrays is in maintaining the temperature of radiating heterojunctions within a narrow range to ensure frequency stability of the radiation. Such array structures consist of a large number of laser diodes soldered to the surface of a perfectly prepared metal mirror at a high packing density on the radiating elements. As the array radiates high-intensity laser light (even at today's demonstrated efficiencies of greater than 60 %), heat exchangers should extract heat flows from the active medium of greater than 1,000 W/cm^2. At this level, the displacement of the radiation spectrum conditioned by a thermal increase of the radiating layer should not increase by more than 3 nm relative to the initial lasing wavelength, which corresponds to a change in temperature of the active layer of not more than 10 °C. The heat exchanger for such a device should have relatively low thermal resistance of not more than 0.1 °C/W.

To obtain these high values of heat removal in devices, high-thermal-conductivity materials such as beryllium ceramics or diamond (with thermal conductivities of 3.7 and 20 W/cm °C respectively) can be used. Unfortunately, the effort and cost to produce and treat these materials makes the process of array fabrication more difficult and expensive. Silicon carbide is frequently used as a thermally conductive material that can be polished to high quality. Besides its high thermal conductivity (in the best case close to that of copper), SiC has high electrical impedance, is environmentally safe and has high hardness—important for optical polishing. Both separate and combined heat-removal elements can be made from SiC as well as complete microchanneled or porous heat exchangers. The coefficient of thermal expansion of SiC is close to that of gallium arsenide, which is the basis of many laser-diode compositions. This similarity helps prevent the material from cracking during heat cycling caused by soldering. Optical-grade SiC is available in sizes up to 1 m in diameter.

Considerable effort has been focused on the development of efficient heat-sinking systems for one-dimensional and two-dimensional laser-diode arrays and improvement of the soldering technology for linear arrays. Research continues into phase-locked arrays of laser diodes, phase-locked systems with efficient injection of radiation into an optical fibre, new configurations for solid-state lasers such as disk and slab lasers and efficient beam steering of high-power laser beams. The development of scalable phase-synchronized two-dimensional arrays of laser diodes emitting high levels of radiation has been made possible in part as a result of advances in the field of high-power optics.

Moscow Victor V. Apollonov

Acknowledgments

The author expresses his gratitude to V.I. Borodin, G.V. Vdovin, Yu.P. Voinov, V.F. Goncharov, S.I. Derzhavin, S.I. Zienko, V.V. Kuzminov, V.N. Motorin, I.G. Odnorozhenko, V.V. Ostanin, S.A. Savranskii, A.G. Safronov, V.Yu. Khomich, S.A. Chetkin, V.A. Shmakov, S.F. Sholev, A.N. Shults and V.A. Shurygin. During the work in the field of power optics and important new applications the author of the book was the scientific supervisor and scientific advisor of separate studies conducted in conjunction with the above researchers.

The author also believes it important to note the fruitful cooperation with V.A. Alekseev, V.N. Antsiferov, D.V. Artemov, G.I. Babayants, B.F. Belov, B.M. Belousov, Yu.A. Broval'skii, F.V. Bunkin, S.V. Bilibin, P.I. Bystrov, M.V. Gartman, Yu.V. Danchenko, L.N. Zakhar'ev, A.A. Kazakov, N.V. Karlov, B.I. Katorgin, V.V. Kiyko, V.I. Kislov, B.Sh. Kishmakhov, V.K. Konyukhov, A.P. Kunevich, N.P. Laverov, V.N. Lugovoi, V.N. Lukanin, E.A. Narusbek, L.M. Ostrovskaya, N.V. Pletnev, N.N. Polyashev, V.N. Rodin, A.P. Sitnov, V.P. Smekalin, Yu.P. Koval', V.N. Kharchenko, E.V. Khristyan, V.N. Timoshkin, V.N.Tishchenko, M.I. Tsypin, A.V. Shirkov and A.A. Shubin who made significant contributions to the development of high-power optics and new important applications.

Contents

About the Author

Victor V. Apollonov (08/07/1945), Doctor of Physics and Mathematics, Professor, Academician of RANS and AES, is a leading specialist in the area of basic principles of creation and development of high-power laser systems and high-power radiation interaction with matter. He is the member of the European and American Physical Society, SPIE, AIAA, American Society for QE and the member of the specialized scientific council of Russia. He is a full member of Russian Academy of Natural Science and Academy of Engineering Sciences, laureate of State Prize of USSR (1982) and of Russia (2001). E-mail: vapollo@kapella.gpi.ru

Part I
Power Optics Elements (POE) for High Power Lasers

The advent of the laser has placed stringent requirements on the fabrication, performance and quality of optical elements employed within systems for most practical applications. Their high power performance is generally governed by three distinct steps, first the absorption of incident optical radiation (governed primarily by various absorption mechanisms); second, followed by a temperature increase and response governed primarily by thermal properties, and finally the elements thermo-optical and thermomechanical response, e.g., distortion, stress birefringenous fracture, etc. All of this needs to be understood in the design of efficient, compact, reliable and useful for many applications high-power systems, under a variety of operating conditions, pulsed, continuous wave, rep-rated or burst mode of varying duty cycles. The second part of the book is devoted to the problem of high-power high repetition rate pulse-periodic laser systems, which in the nearest time will find a lot of applications in the field of space ecology, long range energy transfer, heavy machinery, space engineering, nuclear technologies and many others. The book has considered in detail a new approach to the problem of a laser jet engine creation, which is based on the resonance merging of shock waves generated by an optical pulsating discharge, produced by such lasers. To obtain an optical pulsating discharge, we suggested the usage of high-power pulse-periodic laser radiation, which can be generated by wide aperture carbon dioxide, chemical and mono-module disk-type solid-state laser systems. Future developments of the disk laser technology as the most effective and scalable to the level of many hundreds of kW as well are under consideration in the book.

Chapter 1
Introduction

On the basis of the theory we developed in the early 1970s, a broad range of phenomena is considered in the book for an optical surface of a solid body that is exposed to radiation arbitrarily varying in time and producing temperature gradients, thermoelastic stresses and thermal deformations on the surface layer. The examination is based on the relations between the quantities characterising the thermal stress state in any non-stationary regimes of energy input into a solid, which are similar to Duhamel's integral formula from the theory of heat conduction. A peculiar feature of the analysis of the thermal stress state in this case consists in the fact that this relation comprises time as a parameter, which in turn is a consequence of incoherence of the quasi-stationary problem of thermoelasticity. This phenomenon is particularly important for the optics of high-power, high-pulse repetition rate lasers, which are being actively developed. In the review, we have recently published in Laser Physics, the thermal stress state of a solid is analysed. In this state, time is treated as an independent variable used in differentiation. Such an approach greatly reduces the possibility of the method. The review published contains data on the use of capillary porous structures made of various materials with different degrees of the surface development. Moreover, such structures can be efficiently employed to increase the heat exchange at a temperature below the boiling point of the coolant. In the book we discuss the dependences of the limiting laser intensities on the duration of a pulse or a pulse train, corresponding to the three stages of the state of the reflecting surface and leading to unacceptable elastic deformations of the surface, to the plastic yield of the material accompanied by the formation of residual stresses and to the melting of the surface layer. We also analyse the problem of heat exchange in the surface layer with a liquid metal coolant pumped through it. The theoretical estimates are compared with the experimental data. Book represents the issues related to the technology of fabrication of power optics elements based on materials with a porous structure, of lightweight highly stable large optics based on highly porous materials, multilayer honeycomb structures and silicon carbide, as well as problems of application of physical and technical fundamentals of power optics in modern high-end technology.

V.V. Apollonov, *High-Power Optics*, Springer Series in Optical Sciences 192,
DOI 10.1007/978-3-319-10753-0_1

The second part of the book is devoted to the problem of high power high repetition rate lasers pulse periodic laser systems, which in the nearest time will find a lot of applications in the field of space ecology, long range energy transfer, heavy machinery, space engineering, nuclear technologies and many others. Book has considered in details a new approach to the problem of a laser jet engine creation, which is based on the resonance merging of shock waves generated by an optical pulsating discharge, produced by such a lasers. To obtain an optical pulsating discharge, we suggested the usage of high-power pulse-periodic laser radiation, which can be generated by wide aperture carbon dioxide, chemical and mono-module disk type solid-state laser systems. Future developments of the disk laser technology as the most effective and scalable to the level of many hundreds of kW as well are under consideration in the book.

The history of power optics is inextricably associated with the creation of a single-mode CO_2 laser ($P = 1.2$ kW), operating in the master oscillator–power amplifier regime and employing the principle of a quasi-optical transmission line, at the Laboratory of Oscillations of the P.N. Lebedev Physics Institute headed at the time by A.M. Prokhorov. Its creator was A.I. Barchukov, who worked with a team of young scientists on the problem of scaling of single-mode electric-discharge laser systems [1–5]. Due to the research conducted on such a laser system, we managed to study many physical phenomena occurring when high intensity radiation interacts with matter, including with the elements of the optical path of laser systems, which subsequently greatly facilitated creation of high-power lasers. Then, in the early 1970s, we paid attention to a phenomenon that was to limit undoubtedly the further growth of the power generated by lasers being developed [6]. More than 20 years of fundamental and applied research devoted to the study of this phenomenon and to the solution of problems associated with it allow a conclusion that its essence consists in the following. An optical surface of a highly reflecting power optics element (POE) or any element of an optical path does not fully reflect radiation falling on it. A small portion of energy (fractions of a percent, depending on the wavelength) is absorbed by this reflecting element and turns into heat. As the output power increases, even a small amount of it is sufficient to induce thermal stresses in a POE. Thermal stresses distort the geometry of the reflecting surface, affecting thereby, for example, the possibility of long-distance delivery of radiation and its concentration in a small volume. The discovered effect of thermal deformations of a POE required a theoretical study of the problem that had not been solved in such a setting ever before. Very useful was the experience in solving the problems of thermoelasticity, gained by the theoretical department headed at that time at by B.L. Indenbom at the Institute of Crystallography, USSR Academy of Sciences. Minimisation of the thermoelastic response of the optical surface of the POE exposed to intense laser radiation is one of the key problems of power optics. Improving the efficiency of laser systems, increasing the output power and imposing stricter requirements to the directivity of generated radiation fluxes are inextricably linked with the need to design and create a POE with elastic distortions $\lambda/10 \div \lambda/20$ (λ is the wavelength) at specific radiation loads up to several tens of $kW{\cdot}cm^{-2}$ [7–10].

Interest in power optics and its physical, technical and technological solutions is unabated to this day. An almost simultaneous creation of first lasers in the USA and the USSR gave birth to annual symposia on Optical Materials for High-Power Lasers (Boulder, USA) and Nonresonant Laser–Matter Interaction (Leningrad, USSR). Regular meetings of scientists and engineers, as well as proceedings of the symposia have had a significant impact on the development of research in the field of power optics in many countries [11–13].

The data presented in this book allow one to reconsider important aspects of temperature gradients, thermoelastic stresses and thermal deformations in POEs, resulting from the exposure of their surfaces to high-power laser radiation. In this case, use is made of the relations between the quantities characterising the thermal stress state in any nonstationary regimes of energy input into a solid, which are similar to Duhamel's integral formula from the theory of heat conduction. A peculiar feature of the analysis of the thermal stress state in this case consists in the fact that these relations comprise time t not as an independent variable, which is used in the differentiation (as, for example, in review [14]) but as a parameter, which is a consequence of incoherence of the quasi-stationary problem of thermoelasticity presented below. Thus, by using the theory we developed in the early 1970s, we consider in this review a wide range of phenomena related to the thermal stress state of a solid-body surface exposed to radiation arbitrarily varying in time [15–21]. This consideration is particularly important for the optics of high-power, high-pulse repetition rate laser systems that are being actively developed. In review [14] we analysed important for the development of high-power optics problems of using capillary-porous structures with different degrees of development for the enhancement of heat transfer surface with water temperatures below the boiling point. The review published [14] contains data (important for the development of high-power optics) on the use of capillary porous structures with a different degree of the surface development, which can be efficiently employed to increase the heat exchange at a temperature below the boiling point of the coolant. The evaporation–condensation mechanism of heat transfer in the POE on the basis of porous structures and the idea of lowering the boiling temperature under reduced pressure of the coolant in cellular materials, developed by us at the same time [14, 21], are considered in details.

References

1. A.I. Barchukov, N.V. Karlov, V.K. Konyukhov, Y.B. Konev, B.B. Krynetskii, V.M. Marchenko, Y.N. Petrov, A.M. Prokhorov, D.V. Skobel'tsyn, A.V. Shirkov, FIAN Report, Moscow (1968–1970)
2. A.I. Barchukov, Y.B. Konev, A.M. Prokhorov, Radiotekh. Electron. **15**, 2193 (1970)
3. A.I. Barchukov, Y.B. Konev, A.M. Prokhorov, Radiotekh. Electron. **16**, 996 (1971)
4. V.V. Apollonov, A.I. Barchukov, V.K. Konyukhov, A.M. Prokhorov, FIAN Report, Moscow (1971)
5. V.V. Apollonov, A.I. Barchukov, V.K. Konyukhov, A.M. Prokhorov, FIAN Report, Moscow (1972)

6. V.V. Apollonov, A.I. Barchukov, V.K. Konyukhov, A.M. Prokhorov, Pisma. Zh. Eksp. Teor. Fiz. **15**, 248 (1972)
7. V.V. Apollonov, A.I. Barchukov, A.M. Prokhorov, in *Proceedings on First European Conference on Lasers and Applications*, Drezden, GDR, 1972
8. V.V. Apollonov, A.I. Barchukov, V.K. Konyukhov, A.M. Prokhorov, Sov. J. Quantum Electron. **3**(3), 244 (1973)
9. V.V. Apollonov, A.I. Barchukov, A.M. Prokhorov, in *Proceedings on Second European Conference in Lasers and Applications*, Drezden, GDR, 1973; Preprint FIAN No. 157, Moscow, 1973
10. V.V. Apollonov, A.I. Barchukov, A.A. Lubin, A.M. Prokhorov, A.V. Shirkov, FIAN Report, Moscow (1973)
11. A.J. Glass, A.H. Guenther, Appl. Opt. **12**, 34 (1973)
12. S.J. Cytron, Memorandum Report M73-17-1, Philadelphia, PA (1973)
13. D.H. Jacobson, W. Bickford, J. Kidd, R. Barthelemy, R.H. Bloomer AIAA Paper No 75-719 (1975)
14. V.V. Apollonov, Laser Phys. **23**, 1 (2013)
15. V.V. Apollonov, A.I. Barchukov, A.M. Prokhorov, Radiotekh. Electron. **19**, 204 (1974)
16. V.V. Apollonov, A.I. Barchukov, A.M. Prokhorov, IEEE J. Quantum Electron. **10**, 505 (1974)
17. V.V. Apollonov, E.M. Shefter, Preprint FIAN No 105, Moscow (1974)
18. A.I. Barchukov, Doctoral Dissertation, Moscow, FIAN, 1974
19. V.V. Apollonov, A.I. Barchukov, N.V. Karlov, A.M. Prokhorov, E.M. Shefter, Sov. J. Quantum Electron. **5**, 216 (1975)
20. V.V. Apollonov, Candian Dissertation, Moscow, FIAN, 1975
21. V.V. Apollonov, Doctoral Dissertation, Moscow, FIAN, 1982

Chapter 2
Static POEs Based on Monolithic Materials

Consider the most important aspects of the problem of static POE fabrication, namely, the conditions needed to achieve high optical efficiency thresholds for a mirror surface. Note that in our first studies [1–6] we obtained only stationary expression for the limiting intensities, leading to the optical destruction of POEs, and the stability parameters of optical surfaces based on them.

2.1 Thermal Stress State of a Body Exposed to Laser Radiation

Temperature field We considered a strongly absorbing isotropic body, which at the initial moment of time has a fixed temperature. The body surface with the absorption coefficient A is exposed to an axisymmetric radiation flux of arbitrary temporal shape. It is assumed that the intensity distribution in the laser beam cross section obeys the Gaussian law: $I(\vec{r}) = I_0 \exp(-K_0 r^2)$, where $K_0 = 2/r_0^2$. Energy absorption takes place directly on the irradiated surface. Physically, this means that the skin-layer depth δ is smaller than the depth of the temperature field penetration in the body under consideration during the characteristic times τ of changes in the radiation intensity, i.e., $\delta \ll \sqrt{a^2\tau}$, where a^2 is the thermal diffusivity of the material.

The problem of determining the temperature field was considered in the linear formulation: it was assumed that all thermal and mechanical characteristics of the materials are independent of temperature and energy loss by radiation and convection was neglected. Provided that the characteristic size of the beam is $r_0 < L$, where L is the characteristic size of the irradiated body, and the energy input time is $t < L^2/a^2$, in solving this problem one can use the half-space model. The heating of the sample material is described in this case by the heat conduction equation [7]

$$\frac{\partial T}{\partial t} = a^2 \Delta T \tag{2.1}$$

V.V. Apollonov, *High-Power Optics*, Springer Series in Optical Sciences 192,
DOI 10.1007/978-3-319-10753-0_2

at the following initial and boundary conditions:

$$\left.\frac{\partial T}{\partial z}\right|_{z=0} = -\frac{I_0 A}{\lambda} f(t) \exp(-K_0 r^2),$$
$$T(r, z, 0) = 0, \qquad (2.2)$$
$$\lim_{r,z\to\infty} T < M,$$

where M is the finite quantity; $f(t)$ is the time function of the laser beam intensity normalised to I_0; A_0 is the absorption coefficient of laser radiation on a metal surface; λ is the thermal conductivity of the body material; and T is the temperature.

Using the method of successive integral Hankel and Laplace transforms, we obtain the solution to (2.1):

$$T(r, z, t) = T^* \frac{I_0 A}{2\lambda\sqrt{K_0}},$$
$$T^* = \frac{1}{2\pi i\sqrt{K_0}} \int_{\sigma-i\infty}^{\sigma+i\infty} dp\Psi(p)\exp(pt)$$
$$\times \int_0^\infty \xi \frac{\exp(-\xi^2/4K_0)}{\gamma} \exp(-yz) J_0(\xi r) d\xi, \qquad (2.3)$$

where p and ξ are the parameters of Laplace and Hankel transforms; $\gamma^2 = p/a^2 + \xi^2$; $\Psi(p)$ is the image of the Laplace transform of $f(t)$; and J_0 is the zero-order Bessel function.

This expression allows us to describe the thermal state of a solid body heated by laser radiation, whose intensity varies with time in an arbitrary manner.

Thermoelastic stresses The thermoelastic behaviour of the body is analysed by using the system of equations [7, 8]:

$$\mu\nabla^2\vec{u} + (\lambda' + \mu) grad\, div\,\vec{u} - (3\lambda' + 2\mu)\alpha_T \nabla T + \vec{F} - \rho\ddot{\vec{u}} = 0,$$
$$\nabla^2 T - \frac{1}{a^2}\frac{\partial T}{\partial t} + \frac{W}{\lambda} - \frac{(3\lambda' + 2\mu)\alpha_T T}{\lambda} div\,\vec{u} = 0, \qquad (2.4)$$

where λ' and μ are the Lam e' coefficients [9]; $\vec{u}$ is the deformation vector; ρ is the density of the material; $\vec{F}$ is the external force; α_T is the coefficient of thermal expansion; and W_0 is the density of volume heat sources.

In considering the deformation of an elastic metal half-space whose surface is exposed to pulsed laser radiation, when the inequalities

$$|\rho\ddot{\vec{u}}| \ll (3\lambda' + 2\mu)\alpha_T |\nabla T|,$$
$$\nabla^2 T \sim \frac{1}{a^2}\frac{\partial T}{\partial t} \gg \frac{(3\lambda' + 2\mu)\alpha_T T}{\lambda} div\,\vec{u} \qquad (2.5)$$

are fulfilled, we can pass to the system of equations of the quasi-stationary thermoelasticity theory:

$$\mu\nabla^2\vec{u} + (\lambda' + \mu) grad\, div\, \vec{u} - (3\lambda' + 2\mu)\alpha_T \nabla T = 0,$$
$$\nabla^2 T - \frac{1}{a^2}\frac{\partial T}{\partial t} = 0. \tag{2.6}$$

In this case, from the first inequality we obtain the duration of a single pulse

$$\tau \gg \max\left(\frac{\rho a^2}{\lambda'}; \frac{\rho a^2}{\mu}\right) \sim 10^{-6} \div 10^{-8}\,\mathrm{s}, \tag{2.7a}$$

and from the second

$$\tau^{3/2} \ll \frac{\rho^2 c^2 \sqrt{a^2}}{\mu \alpha_T^2 I_0 A}. \tag{2.7b}$$

We represented the stress tensor components in the general form [10]:

$$\sigma_{zz}^{\wedge} = 2GD\int_0^\infty \xi^2 J_0(\xi r)\varphi(\xi)\left[e^{-\gamma z} - e^{-\xi z}(1 + z(\xi - \gamma))\right]d\xi,$$

$$\sigma_{rr}^{\wedge} = 2GD\int_0^\infty \varphi(\xi)\{J_0(\xi r)[\xi(\xi z - 2)(\xi - \gamma)e^{-\xi z} + \xi^2 e^{-\xi z}$$
$$- \gamma^2 e^{-\gamma z}] + \frac{J_1(\xi r)}{r}\left[\xi e^{-\gamma z} - ((\xi - \gamma)(\xi z - 2(1 - \nu)) + \xi)e^{-\xi z}\right]\}d\xi,$$

$$\sigma_{rz}^{\wedge} = 2GD\int_0^\infty \xi\varphi(\xi)J_1(\xi r)\left[\gamma\left(e^{-\gamma z} - e^{-\xi z}\right) - \xi z(\xi - \gamma)e^{-\xi z}\right]d\xi, \tag{2.8}$$

$$\sigma_{\varphi\varphi}^{\wedge} = 2GD\int_0^\infty \varphi(\xi)\{J_0(\xi r)\left[\left(\xi^2 - \gamma^2\right)e^{-\gamma z} - 2\nu\xi(\xi - \gamma)e^{-\xi z}\right]$$
$$+ \frac{J_1(\xi r)}{r}\left[((\xi - \gamma)(\xi z - 2(1 - \nu)) + \xi)e^{-\xi z} - \xi e^{-\gamma z}\right]\}d\xi,$$

where G is the shear modulus; J_1 is the first-order Bessel function;

$$D = \frac{\alpha_T}{2}\frac{1+\nu}{1-\nu}\frac{I_0 A a^2}{K_0 \lambda p}\Psi(p);\ \ \varphi(\xi) = \frac{\xi}{\gamma}\exp(-\xi^2/4K_0);$$

and ν is Poisson's ratio. Analysis of the expression reveals the nature of the time changes at any point in the half-space.

Thermal deformations The stress state occurring in a solid body is accompanied by its deformation, its largest amplitude being achieved on the irradiated surface. The expression for the normal displacement of the surface, corresponding to a given temperature distribution, has the form:

$$W(r,z,t) = W^* \frac{(1+\nu)\alpha_T I_0 A}{\lambda K_0}, \tag{2.9}$$

$$\begin{aligned} W^* = \frac{F_0}{2\pi i} \int_0^\infty dv \int_{\sigma-i\infty}^{\sigma+i\infty} dp \frac{\Psi(p/t)}{p} \exp(p-v) \\ \times J_0\left(\sqrt{v}\delta_r\right) \frac{\sqrt{v} - \sqrt{v + p/F_0}}{\sqrt{v + p/F_0}}, \end{aligned} \tag{2.10}$$

where $F_0 = 4K_0 a^2 \tau$. The resulting expression allows us to trace the changes in the surface shape during irradiation.

Thus, this consideration has made it possible to describe fully the characteristics of temperature fields, thermoelastic stresses and thermal deformations occurring in solids whose surface is exposed to high-power laser radiation varying with time in an arbitrary manner. In addition, the following relations are fulfilled between the quantities characterising the thermal stress state in the continuous-wave and any other nonstationary regime of energy input into the solid [7, 9]:

$$\begin{aligned} T^{\mathrm{tr}} &= \int_0^t f(t-\tau) \frac{\partial T^{\mathrm{st}}}{\partial \tau} d\tau, \\ \sigma_{ik}^{\mathrm{tr}} &= \int_0^t f(t-\tau) \frac{\partial \sigma_{ik}^{\mathrm{st}}}{\partial \tau} d\tau, \\ W^{\mathrm{tr}} &= \int_0^t f(t-\tau) \frac{\partial W^{\mathrm{st}}}{\partial \tau} d\tau. \end{aligned} \tag{2.11}$$

These relations are similar to Duhamel's integral formula from the theory of heat conduction. It should be noted that the local deformation of the POE surface is the determining factor of the laser impact and the bending deformation component of the POE as a whole can be reduced to zero due to the large thickness of its effectively cooled base. Later, both components of the POE deformation were examined in the book of Tsesnek et al. [11].

2.2 Continuous-Wave Irradiation

Temperature field If the time of laser irradiation satisfies the inequality $r_0^2/a^2 \le t \le L^2/a^2$, a steady-state temperature field can be established in the sample material. The main property of the process of its establishment is described by the expression [10]

$$T^* = \frac{2}{\sqrt{\pi}} \operatorname{arctg}\left(\sqrt{F_0}\right). \tag{2.12}$$

It follows from (2.12) that for instants of times t, at which $F_0 \ge 4$, the current temperature is 10 % less than the steady-state value. We therefore assume that, starting at time t, at which $F_0 > 4$, a stationary thermal state is established in the sample material (Fig. 2.1).

The expression for the temperature field in the half-space has the form [10]

$$T^* = \int_0^\infty J_0\left(\sqrt{v}\delta_r\right) \exp\left[-\sqrt{v}\left(\delta_z + \sqrt{v}\right)\right] \frac{dv}{\sqrt{v}}, \tag{2.13}$$

where $\delta_z = 2\sqrt{K_0}z$ and $\delta_r = 2\sqrt{K_0}r$. From this expression we obtain the locality of the temperature field, the characteristic values of which decrease with increasing distance from the centre of the surface irradiation region and inside the material (Figs. 2.2, 2.3 and 2.4).

Thermoelastic stresses In the steady-state regime ($p \to 0$), nonzero are only the components of the tensor of thermal stresses σ_{rr}^* and $\sigma_{\varphi\varphi}^*$ [10]:

$$\begin{aligned} \sigma_{rr}^* &= 2(1-\nu) \int_0^\infty \exp\left[-\sqrt{v}\left(\sqrt{v} + \delta_z\right)\right] \left[J_1\left(\sqrt{v}\delta_r\right)/\sqrt{v}\delta_r - J_0\left(\sqrt{v}\delta_r\right)\right] \frac{dv}{\sqrt{v}}, \\ \sigma_{\varphi\varphi}^* &= 2(1-\nu) \int_0^\infty \exp\left[-\sqrt{v}\left(\sqrt{v} + \delta_z\right)\right] \left[-J_1\left(\sqrt{v}\delta_r\right)/(v\delta_r)\right] dv, \end{aligned} \tag{2.14}$$

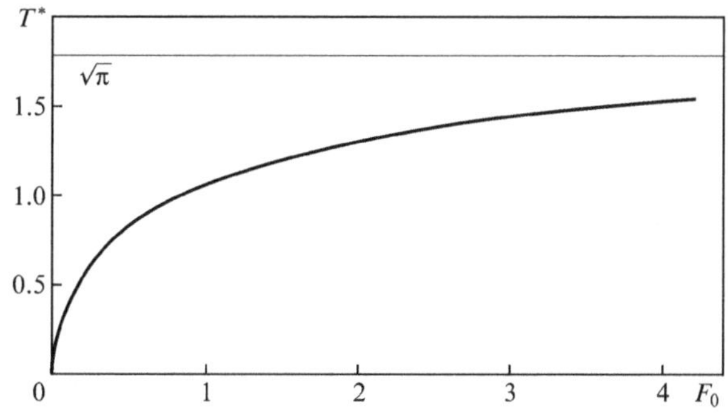

Fig. 2.1 Time dependence of the sample surface temperature at the centre of the region (number F_0) exposed to cw radiation

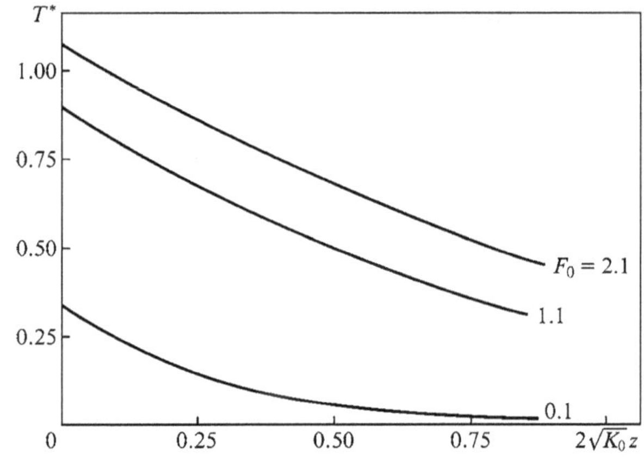

Fig. 2.2 Temperature field distribution on the z axis

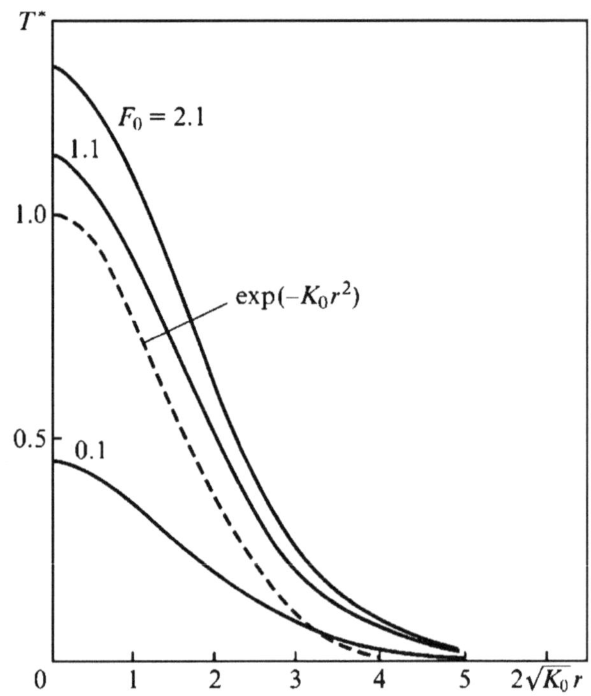

Fig. 2.3 Temperature field distribution on the sample surface

where

$$\sigma_{ik}^{*} = \frac{\lambda\sqrt{K_0}(1-\nu)}{I_0 A G \alpha_T (1+\nu)} \sigma_{ik}(\vec{r}).$$

The maximum values of these components are achieved in the centre of the irradiated region (Fig. 2.5) on the surface of the half-space, where the stationary field of thermoelastic stresses have the form (Figs. 2.6 and 2.7)

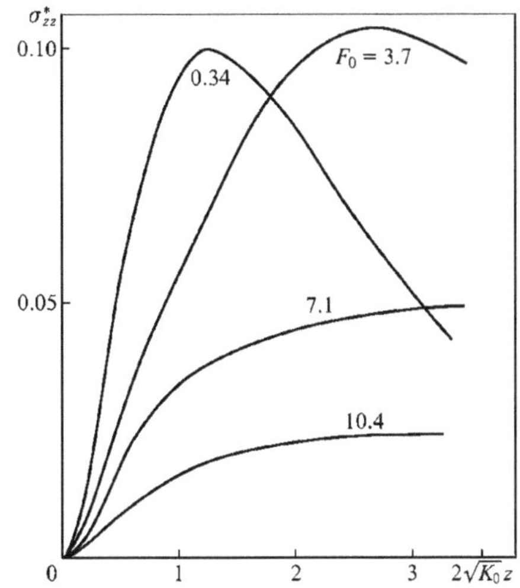

Fig. 2.4 Dependence of the axial stress σ_{zz} on the exposure time of laser irradiation

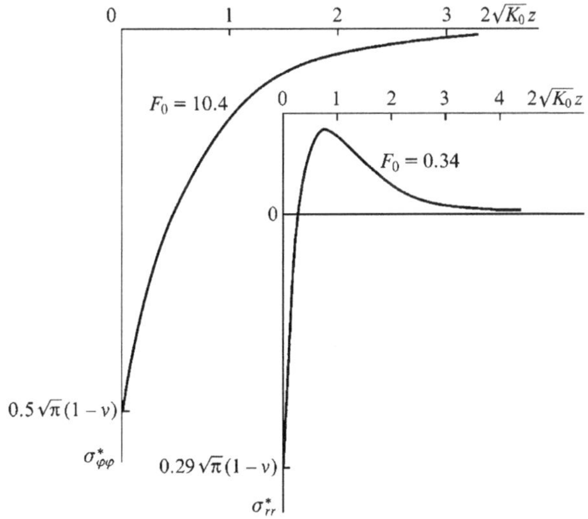

Fig. 2.5 Distribution of the peripheral ($\sigma_{\varphi\varphi}$) and radial (σ_{rr}) tensor components on the z axis for different exposure times of laser irradiation

$$\sigma^*_{rr} = \frac{\sqrt{\pi}(1-\nu)}{2} {}_1F_1\left(\frac{1}{2}; 2; -\delta_r^2/4\right),$$
$$\sigma^*_{\varphi\varphi} = \frac{\sqrt{\pi}(1-\nu)}{2}\left[{}_1F_1\left(\frac{1}{2}; 2; -\delta_r^2/4\right) - {}_1F_1\left(\frac{1}{2}; 1; -\delta_r^2/4\right)\right]. \tag{2.15}$$

The main property in establishing a steady state for σ_{rr} and $\sigma_{\varphi\varphi}$ are characterised by the dependence shown in Fig. 2.8:

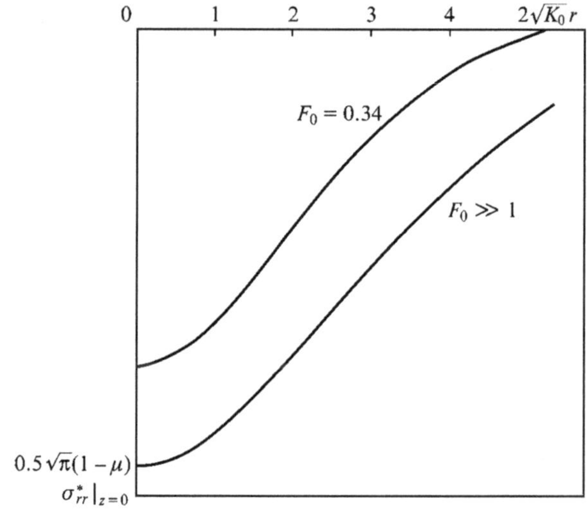

Fig. 2.6 Stress field σ_{rr} on the surface of the half-space

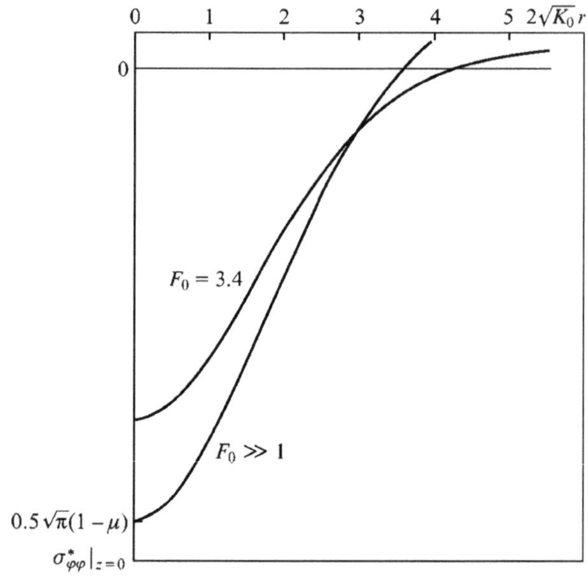

Fig. 2.7 Stress field $\sigma_{\varphi\varphi}$ on the surface of the half-space

$$\sigma^*_{ii}(\delta_r = \delta_z = 0) = \frac{1+\nu}{\sqrt{\pi}}\left[F_0\left(\arctan\frac{1}{\sqrt{F_0}} - \frac{1}{\sqrt{F_0}}\right) - \frac{1-\nu}{1+\nu}\arctan\sqrt{F_0}\right]. \tag{2.16}$$

This expression completely describes the characteristics of the stressed state arising in a solid when its surface is irradiated by cw laser radiation.

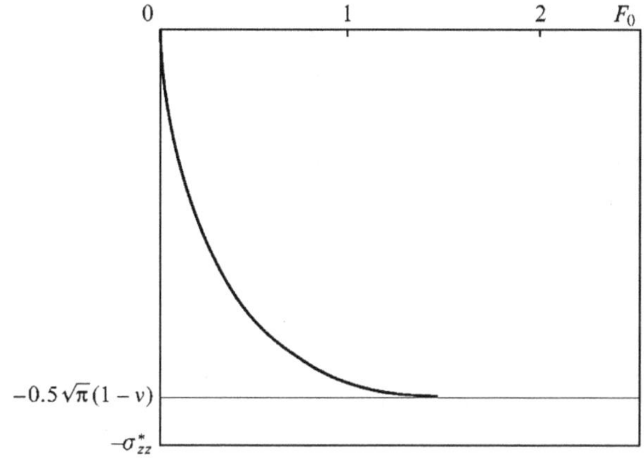

Fig. 2.8 Establishment of a stationary strain state on the surface, in the centre of the irradiated region

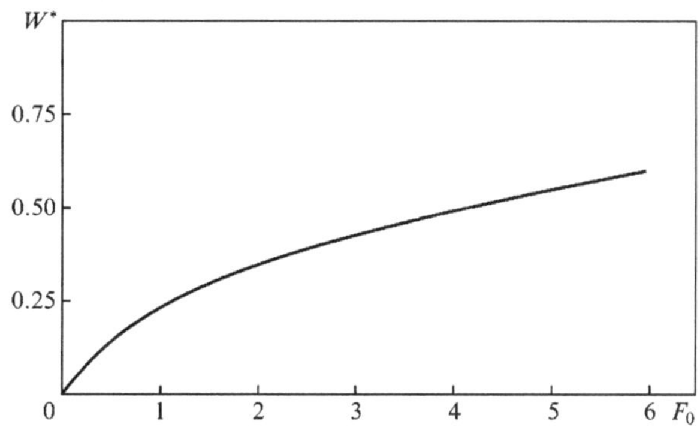

Fig. 2.9 Establishment of a quasi-stationary deformation state on the surface, in the centre of the irradiated region

Thermal deformation of the surface The expression for the displacement W^* of the reflective surface in the half-space model has the form [10]:

$$W^* = -\frac{1}{2}\left\{F_0 \exp\left(-\delta_r^2/4\right) - \left[\frac{4\sqrt{F_0}}{1+F_0} - 2\ln\left(\sqrt{F_0} + \sqrt{F_0+1}\right)\right] {}_1F_1\left(\frac{3}{2}; 1; -\frac{\delta_r^2}{4}\right)\right\}. \tag{2.17}$$

Deformation surface profiles for different exposure times are shown in Fig. 2.9.

2.3 Pulsed Irradiation

Temperature field In the case of short irradiation times, the depth of the temperature field penetration into the material is proportional to $\sqrt{a^2 t} \ll r_0$; therefore, the radial heat spreading can be ignored, and the temperature distribution over the surface repeats the laser beam intensity distribution profile [12]:

$$T^* = \frac{2}{\sqrt{\pi}} \left[\Theta(t^*) \text{arctg}\left(\sqrt{F_0 t^*}\right) - \Theta(t^* - 1) \text{arctg}\left(\sqrt{F_0 (t^* - 1)}\right) \right] \exp(-K_0 r^2), \tag{2.18}$$

where $\Theta(t^*)$ is the Heaviside function; $t^* = t/\tau$; and τ is the pulse duration.

Thermoelastic stresses Thermoelastic stresses arising in a solid irradiated by laser light play an important role in the destruction of the optical surface of the POE. Under pulsed irradiation ($F_0 \ll 1$) the expressions for the stress tensor components are given by (2.15), because in this case the propagation of heat in a solid is of quasi-one-dimensional character and the radial heat spreading can be neglected. The depth of penetration of thermal stresses in the material is $\sqrt{a^2 \tau} \ll r_0$, which follows from the form of σ_{ik} on the z axis:

$$\begin{aligned} \sigma_{rr}^*(\delta_r = 0) &= -\frac{2}{\sqrt{\pi}} \sqrt{F_0} \left[\exp(-\delta_z^2/4F_0) - \frac{\sqrt{\pi}\delta_z}{2F_0} \text{erfc}\left(\frac{\delta_z}{2\sqrt{F_0}}\right) \right] \\ &\approx -\frac{8F_0^{3/2}}{\sqrt{\pi}\delta_z^2} \exp\left(-\frac{\delta_z^2}{4F_0}\right), \end{aligned} \tag{2.19}$$

where

$$\frac{\delta_z}{2\sqrt{F_0}} \gg 1$$

The maximum values of the components σ_{rr}^* and $\sigma_{\varphi\varphi}^*$ are achieved on the surface,

$$\sigma_{rr}^* = \sigma_{\varphi\varphi}^* = -2\sqrt{\frac{F_0}{\pi}} \exp\left(-\frac{\delta_r^2}{4}\right), \tag{2.20}$$

i.e., the distribution of the components σ_{rr}^* and $\sigma_{\varphi\varphi}^*$ on the surface repeat the laser beam intensity distribution. The components σ_{rr}^* and $\sigma_{\varphi\varphi}^*$ on the surface $z = 0$ are equal, and the expression for $\sigma_{ii}^*(\delta_r, \delta_z = 0)$ has the form:

$$\sigma_{ii}^* = -\frac{2}{\sqrt{\pi}} \left[\Theta(t^*) \sqrt{F_0 t^*} - \Theta(t^* - 1) \sqrt{F_0 (t^* - 1)} \right]. \tag{2.21}$$

In the case of small irradiation times

$$\begin{aligned} \sigma_{zz}^* &= 2\delta_z F_0 t^* \int_0^\infty V^3 \exp(-V^2 - V\delta_z) dV, \\ \sigma_{rz}^* &= -\frac{\delta_r}{2} \exp(-\delta_r^2/4) \left[F_0 t^* erf\left(\frac{\delta_z}{2\sqrt{F_0 t^*}}\right) + \frac{\sqrt{F_0 t^*}}{\pi} \delta_z \exp\left(-\frac{\delta_z^2}{4F_0 t^*}\right) \right], \end{aligned} \tag{2.22}$$

where V is a transform variable. The difference in signs of the components means that in the case of thermal deformation of the sample by laser radiation, for σ_{zz} tension of a material is realised, whereas for σ_{rz}—compression. The maximum value of σ_{zz} is achieved on the z axis; in this case, $\delta_z^{\max} \approx \sqrt[4]{12}$, i.e., $z_0^{\max} \approx 0.66 r_0$, and $\sigma_{zz}^{*\max} \approx 1.9 F_0$. The component σ_{rz}^* reaches its maximum value at point $r_0^{\max} = \frac{r_0}{2}$ and $z_0^{\max} \approx 2\sqrt{a^2\tau}$:

$$\sigma_{rz}^{*\max} \approx -0.5 F_0. \tag{2.23}$$

A distinctive feature of the behaviour of the σ_{zz}^* component is that if the inequality $F_0 \ll 1$ is fulfilled, the position of its maximum on the z axis is determined by the spatial characteristics of the laser beam rather than the irradiation time. The maximum of this component is achieved by the end of the laser pulse. This feature is explained by the fact that at $F_0 t^* \ll 1$ the region of thermoelastic perturbations lies on the sample surface and localizes in the irradiation region, because heat due to heat conduction has no time to spread over the sample material. In the opposite case, i.e., at $F_0 t^* > 1$, the point of this component maximum is determined from the condition $(\delta z^2/4F_0 t^*) = 1$.

Thermal deformations The expression for the thermal deformation of the reflecting surface irradiated by a rectangular laser pulse, whose duration satisfies the condition $F_0 \ll 1$, has the form [10]:

$$W^* = -\frac{F_0}{2} \exp(-K_0 r^2) \left[\Theta(t^*) t^* - \Theta(t^* - 1)(t^* - 1)\right]. \tag{2.24}$$

The distribution of thermal deformations of the reflecting surface repeats the laser beam intensity distribution (Fig. 2.10), which we used in our method of the dynamic control of the intensity distribution of laser radiation [13].

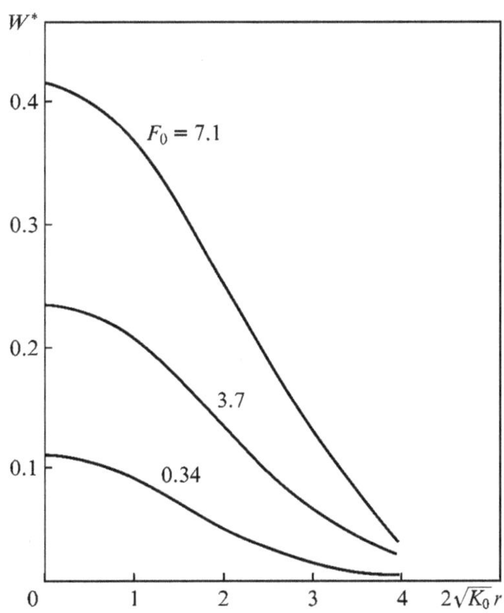

Fig. 2.10 Deformation surface profiles for different exposure times

2.4 Repetitively Pulsed Irradiation

Thermal deformations of a solid body exposed to repetitively pulsed laser radiation were analysed by using the previously derived relations that are similar to Duhamel's integral formulas. The energy flow was treated as a train of rectangular pulses having a duration τ, period T_0 (repetition rate $\nu_0 = 1/T_0$) and off-duty ratio SQ $V = \tau/T_0$. It was assumed that $F_0 = 4K_0a^2T_0 < 1$. The thermal stresses and deformations of the temperature field are expressed in terms of the integrals (typical of the cw regime) that are similar to Duhamel's integrals [14]:

$$F^{PP} = \int_0^t f(t-\tau)\frac{\partial F^{cw}}{\partial \tau}d\tau. \tag{2.25}$$

At the initial instants of time, i.e., when $F_0t^* < 1$, repetitively pulsed irradiation is similar to pulsed irradiation. The geometric meaning of (2.25) is characterised by the area of the integrals in Fig. 2.11. (For the temperature and the components $\sigma_{\varphi\varphi}$ and σ_{rr}, the value of $\frac{\partial F^{cw}}{\partial \tau}$ tends to infinity as $1/\sqrt{t}$ at $t \to 0$ and to zero at $t \to \infty$, for deformation $\frac{\partial F^{cw}}{\partial \tau}$ tends to const at $t \to 0$ and to zero at $t \to \infty$.) In the case of long irradiation times, i.e., when $F_0t^* > 1$, the temperature and thermal stresses reach their quasi-steady states, i.e., a constant component of these values becomes similar to that in the cw regime of energy input with a reduced intensity I_0 SQV. In this case, against the background of this component, along with changes in the laser beam intensity, there will be the characteristic peaks of temperature and stress, which are similar to peaks during pulsed irradiation. A separate 'pulse' of thermal deformations of the reflecting surface exists against the background of a 'stationary component' tending to infinity.

Temperature field The expression for the temperature has the form [10]

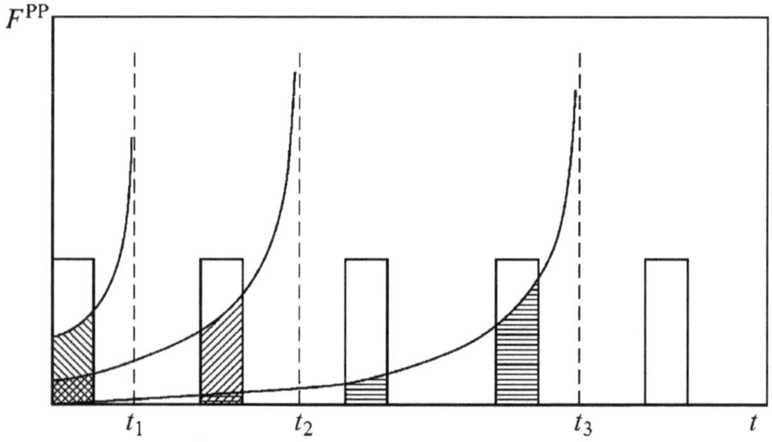

Fig. 2.11 Geometrical interpretation of Duhamel's integrals

$$T^* = \frac{1}{2\pi i} \int_{\sigma_i - i\infty}^{\sigma_i + i\infty} dp \Psi(p) e^p \int_0^\infty \frac{e^{-V} J_0\left(\sqrt{V}\delta_r\right) \exp\left[-\sqrt{V}\left(\sqrt{V} + \delta_z\right)\right]}{\sqrt{8/F_0 + V}} dV, \quad (2.26)$$

since for a train of pulses

$$\Psi(p) = \frac{\left[1 - \exp(-p\tau)\right]\left\{1 - \exp\left[-p(N+1)T\right]\right\}}{p\left[1 - \exp(-pT)\right]} \quad (2.27)$$

is the image of the Laplace transform of $f(t)$; and N is the number of propagated laser pulses.

In the centre of the irradiation region the temperature reaches a maximum value by the time the next pulse terminates ($F_0 > 1$):

$$T^*_{\max} = \sqrt{\pi} SQV + \frac{2}{\sqrt{\pi}} \operatorname{arctg} \sqrt{F_0 SQV}, \quad (2.28)$$

where SQV is the off-duty ration of the temporal structure of radiation.

Thermoelastic stresses Maximum values of the radial and circumferential tangential stress are achieved in the centre of the irradiation region, where they are equal to each other:

$$\sigma_{ii}^{PP} = \int_0^t f(t - \tau) \frac{\partial \sigma_{ii}^{cw}}{\partial \tau} d\tau \quad (2.29)$$

[σ_{ii}^{cw} is determined from (2.16)]. The field distribution of stresses σ_{ii}^* on the surface by the time when the next laser pulse terminates has the form:

$$\sigma_{ii}^* = SQV \sigma_{ii}^{*(1)} + \sum_{n=0}^{N} \Theta(n + 1 - t^*) \left[\Theta(t^* - n)\sigma_{ii}^{*(2)}(t^* - n) - \Theta(t^* - n - SQV)\sigma_{ii}^{*(2)}(t^* - n - SQV)\right], \quad (2.30)$$

where $\sigma_{\varphi\varphi}^{*(1)}$ and $\sigma_{rr}^{*(1)}$ are determined from (2.16), and $\sigma_{rr}^{*(2)} = \sigma_{\varphi\varphi}^{*(2)}$—from (2.21). Because in the steady stress state σ_{zz} and σ_{rz} are identically zero, their values in the case of repetitively pulsed irradiation are the same as in the case of pulsed irradiation (accuracy ~SQV).

Deformation of the surface The displacement of a solid-body surface exposed to repetitively pulsed radiation also has stationary and pulse components [10]: $W^* = SQVW^{(1)*} + W^{(2)*}$.

When the quasi-stationary state is reached

$$W^* = -SQV \ln 2\sqrt{F_0} F_1\left(\frac{3}{2}; 1; -\frac{\delta_r^2}{\pi}\right) - \frac{F_0}{2} \sum_{n=0}^{N} \Theta(n + 1 - t^*) \left[\Theta(t^* - n)(t^* - n) - \Theta(t^* - n - SQV)(t^* - n - SQV)\right] \exp(-K_0 r^2). \quad (2.31)$$

2.5 Criteria for the Optical Surface Stability

Expressions given for the characteristics of the thermal stress state of a solid whose surface is irradiated by high-power cw, pulsed and repetitively pulsed laser radiation allowed us to determine the limiting intensities corresponding to different stages of the optical damage of mirror surfaces [10, 15]. To this end, the parameters of the optical surface stability include not only the thermophysical and mechanical properties of the material but also the parameters of a Gaussian-like beam, namely the intensity in the centre of the irradiation region, the size of the irradiation region and the duration of a single pulse and, in the case of repetitively pulsed irradiation—the pulse train off-duty ratio. The stability parameters of the reflector contain the ratio of a maximum value of the thermal stress state characteristic to its value at which the solid material experiences irreversible macroscopic changes—melting, plastic (brittle) or fatigue deformation or achievement of a critical value $\lambda_0/20$ by the amplitude of thermal deformation of the optical surface, where λ_0 is the wavelength of the laser used. The thus introduced stability parameters of mirrored POE surfaces made it possible not only to compare different pure metals and their alloys in terms of applicability in power optics but also to create specific types of combined POEs capable of withstanding high-power fluxes of cw, pulsed and repetitively pulsed laser radiation.

Continuous-wave regime A solid body whose surface is exposed to cw laser radiation is destroyed when the temperature field in the centre of the irradiation region reaches the melting point of the material and the components of the stress field reach the yield point. The stability of the optical surface under cw irradiation is characterised by the parameters

$$\gamma_{T_{\text{melt}}}^{\text{cw}} = \frac{\sqrt{\pi} I_0 A}{2\lambda\sqrt{K_0} T_{melt}}; \quad \gamma_{\sigma_T}^{\text{cw}} = \frac{\sqrt{3\pi}(1+\nu) I_0 A G \alpha_T}{2\lambda\sqrt{K_0}\sigma_T}. \tag{2.32}$$

If $\gamma_{T_{melt}}$ and $\gamma_{\sigma_T} < 1$, the material will undergo no irreversible changes. The values of these parameters in the case of cw laser radiation at a power density $I_0 A = 1$ kW cm^{-2} and $K_0 = 8 \times 10^2$ m^{-2} are shown in Table 2.1 for Cu, Al and Mo. The main reason for the damage of the optical surface can be determined from the relation

$$\gamma_{\text{rel}}^{\text{cw}} = \frac{\gamma_{\sigma_T}^{\text{cw}}}{\gamma_{T_{\text{melt}}}^{\text{cw}}} = \frac{\sqrt{3}(1+\nu) G \alpha_T T_{\text{melt}}}{\sigma_T}. \tag{2.33}$$

If $\sigma_{\text{rel}}^{\text{cw}} > 1$, the material will be destroyed when the component σ_{ii} reaches the yield point, or when the melting point of the material, $T(0, 0, \infty)$ is reached.

For the materials in question (Table 2.1), the main reason for the deterioration of the optical surface at lower laser intensities is irreversible plastic deformations of the POE in the centre of the irradiation region. There is another important reason for the deterioration of the optical surface—excess of the critical value $\lambda_0/20$

Table 2.1 Parameters of stability and threshold intensities for Al, Mo and Cu at $I_0A = 1\,\text{kW cm}^{-2}$, $r_0 = 5$ cm and $t = 5 \times 10^{-5}$ s

Parameter	Material		
	Cu	Mo	Al
CW regime			
$\gamma_{T_{melt}} = \frac{\sqrt{\pi} I_0 A}{2\lambda\sqrt{K_0}} \frac{1}{T_{melt}}$	0.74	0.8	2.3
$I_{th} = \frac{1}{\gamma_{T_{melt}}}$ /kW cm^{-2}	1.4	1.3	0.44
$\gamma_{\sigma_T} = \frac{\sqrt{3\pi} I_0 AG\alpha_T(1+\nu)}{2\lambda\sqrt{K_0}\sigma_T}$	19.3	10^4	38.3
$I_{th} = \frac{1}{\gamma_{\sigma_T}}$ /kW cm^{-2}	0.05	10^{-4}	2.6×10^{-2}
Pulsed regime			
$\gamma_{T_{melt}} = \frac{2I_0 A\sqrt{a^2\tau}}{\sqrt{\pi}\lambda T_{melt}}$	2.0×10^{-3}	1.45×10^{-3}	5.4×10^{-3}
$I_{th} = \frac{1}{\gamma_{T_{melt}}}$ /kW cm^{-2}	500	700	190
$\gamma_{\sigma_T} = \left(\frac{3a^2\tau}{\pi}\right)^{1/2} \frac{2I_0 AE\alpha_T}{\lambda(1-\nu)\sigma_T}$	0.16	55	0.28
$I_{th} = \frac{1}{\gamma_{\sigma_T}}$ /kW cm^{-2}	6.3	1.8×10^{-2}	3.6
$\gamma_{\lambda_0/20} = \frac{40(1+\nu)I_0 AE\alpha_T a^2\tau}{\lambda\lambda_{CO_2}}$	0.013	4.6×10^{-3}	2.6×10^{-2}
$I_{th} = \frac{1}{\gamma_{\lambda_0/20}}$ /kW cm^{-2}	80	230	39
$\gamma_{\sigma_b} = \frac{4I_0 AE\alpha_T a^2\tau\sqrt{K_0}}{(1+\nu)\lambda_{\sigma_b}}$	2	1.6×10^{-2}	7.4
$I_{th} = \frac{1}{\gamma_{\sigma_T}}$ /kW cm^{-2}	0.5	6.3×10^{-4}	0.14
$\gamma_{\sigma_f}(\sigma_{rz}) = \frac{I_0 AG\alpha_T(1-\nu)F_0}{2\lambda\sqrt{K_0}\sigma_f}$	1.8×10^{-4}	0.04	2.7×10^{-4}
$I_{th} = \frac{1}{\gamma_{\sigma_T}}$ /kW cm^{-2}	5.5×10^3	25	3.7×10^3
$\gamma_{\sigma f}(\sigma_{zz}) = \frac{2I_0 AG\alpha_T(1+\nu)F_0}{\lambda\sqrt{K_0}(1+\nu)\sigma_f}$	1.2×10^{-3}	0.26	1.8×10^{-3}
$I_{th} = \frac{1}{\gamma_{\lambda_0/20}}$ /kW cm^{-2}	830	3.8	560

by the value of thermal deformation of the optical surface—which is implemented at long exposure times of high power laser radiation and in the range of the parameters corresponding to the elastic deformation of the material. In this case, phase and energy characteristics of the reflected laser beam are markedly impaired. The criterion for the optical surface stability to such changes in the optical characteristics of the reflector is given by parameter

$$\gamma^{\mathrm{cw}}_{\lambda_0/20} = \frac{20(1+\nu)\alpha_T I_0 A}{\lambda K_0 \lambda_0} \ln 2\sqrt{F_0}. \tag{2.34}$$

The value $\gamma^{\mathrm{cw}}_{\lambda_0/20} < 1$ can be reached if use is made of some types of reflector designs with efficient cooling [16].

Pulsed regime The parameters of the optical surface stability under pulsed irradiation by a Gaussian-like laser beam having a duration τ and intensity I_0 in the centre of the irradiation region, determined by the ability to reach critical values T_{melt}, σ_T and $\lambda_0/20$ by temperature $T(0,0,\tau)$, thermoelastic stresses $\sigma_{ii}(0,0,\tau)$ and thermal deformations $W(0,0,\tau)$, have the form [13–15]:

$$\begin{aligned}
\gamma^P_{T_{\mathrm{melt}}} &= \frac{2I_0A}{\sqrt{\pi}\lambda T_{\mathrm{melt}}}\sqrt{a^2\tau},\\
\gamma^P_{\sigma_T} &= 4\sqrt{\frac{3}{\pi}}\frac{I_0AG\alpha_T(1+\nu)}{\lambda G_T(1-\nu)}\sqrt{a^2\tau},\\
\gamma^P_{\lambda_0/20} &= \frac{40(1+\nu)I_0A\alpha_T a^2\tau}{\lambda\lambda_0}.
\end{aligned} \tag{2.35}$$

The values of these parameters, found for copper, aluminium and molybdenum at $I_0A = 1$ kW cm^{-2}, $K_0 = 2.82 \times 10^2$ m^{-2} and $\tau = 50$ µs, and the heat flow values I_0A, at which $\gamma^P_i = 1$, are presented in Table 2.1. In the cw regime, the optical surface properties are mainly degraded due to irreversible plastic deformations in the centre of the irradiation region. Under pulsed irradiation the behaviour of the thermal stress state is more complicated than under cw irradiation. Thus, in contrast to the stationary thermal stress state, the nonstationary state in the material of a solid is characterised by the presence of the nonzero components σ_{zz} and σ_{rz}. In this case, the highest value is reached by the component σ_{zz} on the z axis at a distance of $\sim 0.66 r_0$ from the optical surface. If at some level of these I_0A values the component σ_{zz} is greater than the strength σ_b, it is possible to implement the conditions of brittle fracture, at which the surface layer of the POE material will be detached. For this type of destruction the parameter of the optical surface stability has the form:

$$\gamma^u_{\sigma_b} = \frac{4I_0AE\alpha_T a^2\tau\sqrt{K_0}}{(1-\nu)\lambda\sigma_b},$$

and the stability parameter defined with respect to plastic deformation, has the form:

$$\gamma^P_{\sigma_T} = \frac{\sqrt{3}I_0AE\alpha_T F_0}{\lambda\sqrt{K_0}(1-\nu)\sigma_T}\left(1 + 2\frac{\sqrt{F_0}}{3\pi}\exp\left(-\frac{\sqrt{3}}{2F_0}\right)\right). \tag{2.36}$$

The values of the parameters and their corresponding intensities for Al, Mo and Cu are listed in Table 2.1.

Repetitively pulsed regime The state of a solid body, whose surface is irradiated by repetitively pulsed laser pulses, combines the characteristic features of thermal stress states implemented under pulsed and cw irradiation. In this case, for the temperature fields, the fields of the components σ_{rr} and $\sigma_{\varphi\varphi}$, the stress tensor and the thermal deformation fields the realisable temperature and thermal stress states are a combination of stationary and nonstationary states. In this regard, the stability parameters of the reflecting surfaces, defined by the ability of the temperature to reach the melting point of the material, of the components σ_{rr} and $\sigma_{\varphi\varphi}$ to reach the yield point and of thermal deformation to reach the threshold $\lambda_0/20$, are as follows [10]:

$$\gamma_i^{PP} = SQV\gamma_i^{\mathrm{cw}} + \gamma_i^{P}. \tag{2.37}$$

Under repetitively pulsed irradiation, a nonstationary, cyclically repeated stress state arises on a solid surface in the material. As a result, the material of the solid body may experience irreversible fatigue damage. The conditions under which the POE surface undergoes macroscopic fatigue fracture can be assessed by Wöhler curves, determining the dependence of modulus of the amplitude of fatigue stresses on the number of cycles of the loading pulses N_p [10–17].

2.6 Irreversible Changes of the Optical Surface

Dynamics of the fatigue and brittle fracture is characterised by the emergence and extension of microcracks. Therefore, inadmissibility of destruction of the optical POE surface is dictated by the need to preserve the diffusely scattered component of laser radiation at negligible levels. Moreover, the origin and development of microcracks is accompanied by microstructural and phase transformations of the material, leading to a change in the structural and phase composition of the reflecting surface and, as a consequence—to an increase in its absorption coefficient A, whereas the adsorption of various substances on the resulting system of microcracks initiating an optical breakdown leads to a decrease in radiation resistance of the reflecting surface. Furthermore, the optical breakdown of air near the target can occur without the segregation of impurities directly in the vicinity of emergence of microcracks, because they become the nucleus of the electric fields, etc. We considered sequentially the basic mechanisms of microstructural and phase transformations preceding the stage of plastic, fatigue and brittle fracture or accompanying these stages, as well as analysed the possible reasons for the change in the optical surface quality. The expressions obtained are important not only for the problems of power optics. They are effectively used today for the analysis of the conditions of fracture of solids of different nature due to excess of limiting stresses for the various components of the stress tensor.

References

1. V.V. Apollonov, A.I. Barchukov, V.K. Konyukhov, A.M. Prokhorov, FIAN Report, Moscow (1971)
2. V.V. Apollonov, A.I. Barchukov, V.K. Konyukhov, A.M. Prokhorov, FIAN Report, Moscow (1972)

3. V.V. Apollonov, A.I. Barchukov, V.K. Konyukhov, A.M. Prokhorov, Pis'ma. Zh. Eksp. Teor. Fiz. **15**, 248 (1972)
4. V.V. Apollonov, A.I. Barchukov, A.M. Prokhorov, in *Proceedings of First European Conference on Lasers and Applications*, Drezden, GDR, 1972
5. V.V. Apollonov, A.I. Barchukov, V.K. Konyukhov, A.M. Prokhorov, Sov. J. Quantum Electron. **3**(3), 244 (1973)
6. V.V. Apollonov, A.I. Barchukov, A.M. Prokhorov, in *Proceedings of Second European Conference on Lasers and Applications*, Drezden, GDR. Preprint FIAN No. 157, Moscow (1973)
7. H. Parkus, *Thermoelasticity* (Fizmatgiz, Moscow, 1963)
8. W. Nowacki, *Problems of Thermoelasticity* (Izd. AN SSSR, Moscow, 1962)
9. E. Jahnke, F. Emde, F. Lusch, *Special Functions* (Nauka, Moscow, 1964)
10. V.V. Apollonov, Doctoral Dissertation, Moscow, FIAN, 1982
11. L.S. Tsesnek, O.V. Sorokin, A.A. Zolotukhin, *Metal mirrors* (Mashinostroenie, Moscow, 1983)
12. V.V. Apollonov, A.I. Barchukov, A.M. Prokhorov, in *Proceedings of V All-Union Meeting On Nonresonant Interaction of Optical Radiation with Matter*, Leningrad, 1977
13. V.V. Apollonov, A.I. Barchukov, L.T. Kir'yanova, L.M. Ostrovskaya, A.M. Prokhorov, V.N. Rodin, O.S. Serebryannikova, M.I. Tsypin, FIAN Report, Moscow (1977)
14. V.V. Apollonov, A.I. Barchukov, L.M. Ostrovskaya, A.M. Prokhorov, V.N. Rodin, M.I. Tsypin, Patent No. 103162
15. V.V. Apollonov, A.I. Barchukov, A.M. Prokhorov, E.V. Khristyan, Lett. JTP **4**, 433 (1978)
16. V.V. Apollonov, A.I. Barchukov, V.I. Borodin, P.I. Bystrov, V.F. Goncharov, A.M. Prokhorov et al., Lett. JTP **4**, 1193 (1978)
17. V.V. Apollonov, A.I. Barchukov, L.M. Ostrovskaya, A.M. Prokhorov, M.I. Rodin Tsypin, Sov. J. Quantum Electron. **5**, 446 (1978)

Chapter 3
Static OPEs Based on Materials with a Porous Structure

The feasibility of using porous structures for cooling thermally stressed POEs was justified theoretically and experimentally in our papers [1–8]. An increase in the optical damage threshold of laser reflectors based on porous structures was provided by a 'minimum' thickness of the separating layer (tens of microns), by the heat transfer intensification, by high permeability of the heat exchanger for the selected coolants pumped through the porous structure and by the use of the heat exchanger with significantly developed surface. The test results of water-cooled POEs that are based on the porous structures indicated the possibility of removal of high heat flows at low values of the mirror surface deformations. The maximum density of the heat flow being removed, which does not lead to destruction of the mirror surface, was equal to 8.2 kW cm^{-2}. At $q = 2$ kW cm^{-2} the value of thermal deformation was $\sim\lambda_0/20$, where $\lambda = 10.6$ μm [1, 2].

A further increase in the optical damage threshold of cooled mirror surfaces can be realised by the optimisation of the porous structure parameters [3, 7, 8], the appropriate choice of the coolant [9], the development of the technology of fabrication a thin separating layer based on intermetallic compounds [1] and the rational design of the POE on the whole [4–6]. The development of a cooled POE requires a detailed study of heat and mass transfer in porous structures. These processes at the beginning of research in the field of power optics were either insufficiently studied or not studied at all.

3.1 Temperature Field in Porous Structures Under Convective Cooling

The temperature fields in porous structures are calculated in the one-dimensional formulation under the following assumptions: the incident radiation is uniformly distributed over the irradiated surface; the thickness of the porous layer Δ_p is much greater than the depth of heating, which makes it infinitely large ($\Delta_p \to \infty$) and allows consideration of the half-space model; and the temperature and velocity

V.V. Apollonov, *High-Power Optics*, Springer Series in Optical Sciences 192,
DOI 10.1007/978-3-319-10753-0_3

of the flow through the thickness the porous layer are constant. The heat transfer equation, which describes the temperature distribution over the thickness of the porous layer, can be written in the dimensionless form:

$$\frac{d^2\Theta}{d\bar{x}^2} = N(\Theta - 1), \tag{3.1}$$

where $\Theta = t/t_T$, $\bar{x} = x/d_m$ and $N = \tilde{N}\mathrm{u}N'$ are the dimensionless temperature, coordinate and Nusselt number, respectively; d_m is the mean diameter of the structure particles; $\tilde{N}\mathrm{u} = h_m d_m/\tilde{\lambda}$ is the modified Nusselt number, which characterises the ratio of the convective cooling to the heat transfer due to skeleton thermal conductivity; $N' = S_V d_m$ is a dimensionless parameter; and S_V is the heat transfer surface.

The boundary conditions of this equation can be written in the form:

$$\bar{x} = 0;\ d\Theta/d\bar{x} = -\tilde{N}\mathrm{u}\,\bar{q}\,\bar{x} \to \infty;\ \Theta \to 1, \tag{3.2}$$

where $\bar{q} = q/(h_m t_T)$ is the dimensionless heat flux density and q is the heat flux density transmitted through the separating layer. The solution to this equation has the form [3, 9]

$$\Theta(\bar{x}) = 1 + \bar{q}\sqrt{\tilde{N}\mathrm{u}/N'}\exp(-\sqrt{N}x). \tag{3.3}$$

It follows from (40) that the rate of temperature decrease over the thickness of the porous structure is determined by the parameter $\sqrt{N}$. The maximum heat flux density, removed from the reflector due to convective cooling, follows from the condition of equality of the coolant temperature Θ_p at a fixed pressure to the boiling temperature Θ_boil of the coolant ($\Theta_\mathrm{p} \approx \Theta_\mathrm{boil}$) at a chosen pressure and has the form:

$$\bar{q}\mathrm{max} = (\Theta_\mathrm{boil} - 1)\sqrt{N'/\tilde{N}\mathrm{u}}. \tag{3.4}$$

The degree of heat transfer intensification in a porous structure as a result of the turbulent flow circulation and the surface development is determined by the coefficient K_int, which characterises the ratio of the amount of heat removed by the coolant in the structure under consideration to the amount of heat that would be removed directly from the cooling surface of the separating layer by the coolant when it flows in a slot gap of depth Δ [10]:

$$K_\mathrm{int} = q/h_\Delta(t_{\bar{x}=0} - t_T), \tag{3.5}$$

where h_Δ is the coefficient of convective heat transfer in the coolant flow in a slot gap. For example, for the turbulent regime of the coolant flow the Nusselt number Nu_Δ has the form

$$\mathrm{Nu}_\Delta = 0.023\,\mathrm{Re}^{0.8}\mathrm{Pr}^{0.4}, \tag{3.6}$$

where Re and Pr are the Reynolds and Prandtl numbers.

In the case of removal of heat fluxes, the depth of heating is [11]

$$\bar{\Delta}_{\max} = N^{-1/2} \ln 10^2(\Theta_{\text{boil}} - 1). \tag{3.7}$$

In combination with the expressions describing the flow hydrodynamics, the obtained dependences are the basis for optimising the parameters of porous structures, ensuring minimal thermal deformations of POE surfaces or, if necessary, the maximum heat fluxes, removed in the case of convective cooling.

3.2 Convective Heat Transfer in a Porous Structure

The regime of the coolant flow in porous materials, which is of interest for high-power optics, is a transition between laminar and turbulent regimes. The criterion equation of the interporous convective heat transfer for gases and droplets can be represented in the form [6–11]

$$\text{Nu} = c(\text{Re Pr})^n, \tag{3.8}$$

where c and n are the constants depending only on the structural characteristics of the porous material.

Using known experimental data from the literature [12], we analysed the dependences of c and n on the structural characteristics of porous structures for which these constants are quite authentically known. As a result, we found that c and n depend mainly on the bulk porosity Π_V. Thus, relation (3.8) for the dimensionless Nusselt number with account for correlation expressions $c(\Pi_V)$ and $n(\Pi_V)$ allows one to calculate the coefficient of convective heat transfer in porous structures.

3.3 Hydrodynamics of a Single-Phase Flow in a Porous Structure

The temperature field and thermal deformation of the POE are largely determined by the flow rate of the coolant pumped through a porous layer, which depends on the hydrodynamic characteristics and conditions of the coolant inlet and outlet. Hydrodynamic characteristics of a single-phase fluid flow in porous structures, mainly in the region $\Pi_V \leq 0.5$, were studied in many experimental papers [11, 12]. In the general case, the hydrodynamics of the flow in porous structures is described by the modified Darcy's equation (Dupuit–Reynolds–Forchheimer equation [(3.6–3.8)]):

$$-\frac{dp_0}{dx} = \alpha\mu_0 u + \beta\rho u^2, \tag{3.9}$$

where p_0 is the flow pressure; u is the filtration rate, equal to the ratio of the specific mass flow rate of the coolant G_0 to the density ρ; α and β are the viscous and inertial resistance coefficients, respectively; and μ_0 is coefficient of dynamic viscosity of the coolant.

From (3.9) we obtained the equation for the coefficient of friction, C_f, in the form

$$C_f = 2/\mathrm{Re} + 2, \tag{3.10}$$

where $C_f = -2(dp/dx)\rho/G^2\beta$ and Re = $G\beta/\mu\alpha$ (the characteristic size β/α).

Known also is a slightly different approach to the calculation of C_f: as a characteristic size use is made of $\sqrt{K}$, where K is the permeability coefficient, characterising the hydrodynamics of the flow according to Darcy's law ($\mathrm{Re}_{\sqrt{K}} = G\sqrt{K}/\mu$), then

$$C_f = 2(1/\mathrm{Re}_{\sqrt{K}} + c)/c. \tag{3.11}$$

The relationship between the coefficients α, β and parameters c and K can be represented as $\alpha = 1/K$ and $\beta = c/\sqrt{K}$. The parameter c is a universal constant for identical porous structures. For example, for all the materials made of metal powders with spherical or close-to-spherical particles $c \approx 0.55$, and for materials made of powders of arbitrary particle shape $0.45 < c < 0.566$. Thus, when calculating the hydraulic characteristics of the structures we assumed $c = 0.55$, although in our case this provides a somewhat higher value of the friction coefficient C_f.

The permeability coefficient K, which is a structural characteristic of a porous structure, does not depend on the flow regime and is determined experimentally from Darcy's law. In connection with the development of works in the field of heat pipes, many experimental data are currently available to determine K for powder and metal fibrous structures. The dependence of the permeability coefficient for metal fibrous structures on bulk porosity has the form [13]:

$$K = A\Pi_V^m, \tag{3.12}$$

where A and m are the coefficients depending on the relative length of the fibers l/d. Similar expressions can be obtained for powder porous structures. In addition, the permeability coefficient is calculated from the known Carman–Kozeny relation:

$$K = \varphi\Pi_V^3 d_s^2/(1 - \Pi_V)^2 \approx \Pi_V^3/5S_V, \tag{3.13}$$

where φ is a constant depending on the structure.

We used the expressions presented to determine the hydraulic characteristics of power optics elements utilising porous structures made of metal powders and metal fibrous structures.

3.4 Effect of the Coolant Inlet and Outlet Conditions on the Hydraulic Characteristics of the POE

Usually, in cooled POEs the coolant is supplied to and removed from the porous structure through evenly distributed alternating channels on the surface being cooled. In the case of inlets and outlets in the form of alternating holes we may deal with a significant nonuniformity of the velocity field in calculating the flow in radial directions [14–16]. This leads to additional pressure drops in the circulation of the coolant, which are accounted for by the coefficient K_g. In this case, the total pressure drop in a porous structure has the form

$$\Delta P_0 = \Delta p_0 K_g, \tag{3.14}$$

where Δp_0 is the pressure drop in the case of a uniform velocity field.

The coefficient K_g, characterising the influence of collector effects on hydraulic resistance during the motion of the fluid in a porous structure, can be written as:

$$K_g = \frac{F}{\pi s^2 n(1-a)} \frac{cG(1/a-1)/\pi n \rho s \Delta - (\nu/\sqrt{K}) \ln a}{\nu/\sqrt{K} + c\nu}. \tag{3.15}$$

Here F is the area of the irradiated surface; n is the number of channels for the inlet (outlet) of the coolant; and $a = 2r_0/s$ is a relative spacing between the holes. One can see from (3.14) that K_g depends both on the geometric characteristics of the supply and removal of the coolant (on a) and on G; with increasing a and G, the coefficient K_g increases. Thus, the coefficient K_g characterises the design excellence of the inlet and outlet system of the POE coolant.

When K_g is known, the total pressure drop in the porous structure is calculated by formula (51), taking into account the expression for calculating Δp_0:

$$\Delta p_0 = \nu \rho s(1-a)(\nu/\sqrt{K} + c\nu)/\sqrt{K}, \tag{3.16}$$

where $v = Cs/(\rho F \Delta)$ is the coolant filtration rate.

3.5 Thermal Conductivity of Porous Structures in POEs

As for the problems of cooling of POEs, of interest is to study the thermal conductivity of a porous structure skeleton. In most cases, data are summarised in the form of the dependence $\tilde{\lambda}(\Pi_V)$ for the samples, manufactured using the single technology and the same type of material. In calculations use can be made of the Odolevsky equation:

$$\tilde{\lambda} = \lambda_c(1 - \Pi_V)/(1 + \Pi_V), \tag{3.17}$$

where λ_c is the thermal conductivity of a compact material.

Effective thermal conductivity of metal fibrous felt structures may have a considerable anisotropy depending on the direction of the fibres in the felt. Usually, $\tilde{\lambda}$ is generalised by the relations [14]:

$$\tilde{\lambda}_{||} = \lambda_c(1 - \Pi_V)\exp(-\Pi_V), \tag{3.18a}$$

$$\tilde{\lambda}_{\perp} = \lambda_c(1 - \Pi_V)^2, \tag{3.18b}$$

where $\tilde{\lambda}_{||,\perp}$ is the effective thermal conductivity in the direction parallel and perpendicular to the felt-making plane. In the latter case one can also use the expression

$$\tilde{\lambda}_{\perp} = \lambda_c(1 - \Pi_V)^3. \tag{3.19}$$

The expressions presented satisfactorily approximate the experimental data [10, 14] and can be used in the determination of the thermal characteristics of cooled POEs made of metal fibrous structures. In this case, expression (3.18b) describes the upper limit of the experimental data (an optimistic estimate), and (3.19)—the lower (a pessimistic estimate).

3.6 Thermal Deformation of POE Optical Surface Based on Porous Structure

To assess small distortions of the optical surface, which are characteristic of POE deformation, we made an assumption of free expansion of the porous structure and the separating layer according to the temperature fields. Then, the thermal deformation of the mirror surface W^* is the sum of expansions of the separating (thickness Δ_s) and porous (thickness Δ) layers:

$$W^* = \alpha_s\Delta_s t_T[1/2(\Theta_1 + \Theta_2) - \Theta_0] + \alpha_p\Delta t_T[(1 - \Theta_0) + \bar{q}d_m/(N'\Delta)], \tag{3.20}$$

where α_s and α_p are the temperature coefficients of linear expansion of the separating and porous layers, respectively; $\Theta_i = t_i/t_T$ is the dimensionless temperature; t_0 is the coolant temperature at the inlet of the reflector; and t_1 and t_2 are the temperatures of the outer and inner surfaces of the separating layer, respectively.

The above-derived expressions describing the processes of heat and mass transfer in porous structures, were used to calculate the characteristics of cooled POEs [2–11].

Figure 3.1 shows the qualitative dependence of thermal deformation on the maximum density of the heat flow being removed (the variable Π_V at a constant d_m) for two selected regions of the reflector, corresponding to the regions of injection $[W_1^*(q_{\lim 1})]$ and outflow $[W_2^*(q_{\lim 2})]$ of the coolant. Varying the average grain size d_m (or the diameter of the fibre), we can construct a family of curves, characterised by a constant value of d_m and variable porosity V, for reflectors with the same type of the capillary structure. The curves were plotted at a constant pressure drop and by

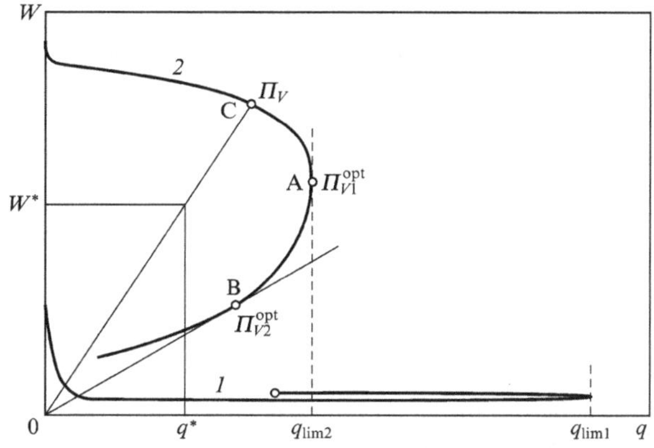

Fig. 3.1 Qualitative dependence of thermal deformations on the maximum power density for the heat flow removed from two POE zones, corresponding to the regions of *1* inlet and *2* outlet of the coolant

taking into account the coolant heating in the porous structure; in addition, the temperature of the coolant at the inlet was assumed equal to the POE temperature.

The deformation of the optical surface in the coolant outflow region is $W_2^* > W_1^*$; hence, crucial to the selection of the structural characteristics of the porous structure is curve (*2*), and the difference between curves (*1*) and (*2*) characterises the degree of perfection of the cooling system. The curves are the envelopes of the working thermal deformation characteristics of a family of reflectors with this type of structure. The performance characteristic of the reflector with a given porosity of the structure Π_V is obtained by connecting the straight line from point *C* with the origin of the coordinates. Point *C* corresponds to the maximum density of the heat flow, removed due to convective cooling, and the line segment *OC* is a dependence of thermal deformations of the mirror surface on the heat load.

In general, curve (*2*) has two points: point A corresponding to optimal porosity Π_{V1}^{opt} which facilitates removal of the heat flow having the maximum density for the selected grain size, the coolant pressure drop, and coolant inlet and outlet conditions; and point B [the point of tangency of curve (*2*) with the straight line from the origin of the coordinates] corresponding to the porosity Π_{V2}^{opt}, for which in the porous structure the optimal thermal distortions of the mirror surface are realised.

The choice of material and the basic parameters of the structure (d_m and Π_V) must be based on a comparison of a family of curves (*2*) with possibilities of obtaining the desired porous structures and separating layers. In our review [17] we have presented the results illustrating the feasibility of the experimental method and numerical calculations of thermal deformation characteristics. The choice of material and the basic parameters of the structure (d_s, Π_V) must be based on a comparison of a family of curves 2 with possibilities of obtaining the desired porous structures and separating layers.

Figures 3.2a–d present the results of numerical calculations of thermal deformation characteristics of a family of water-cooled reflectors with porous structures made of the widely used copper powders (Fig. 3.2a, b) and molybdenum powders (Fig. 3.2c, d), illustrating the capabilities of the method. The average grain size and bulk porosity varied between 20 μm $\leq d_s \leq$ 200 μm and $0.1 \leq \Pi_V \leq 0.9$.

One can see from Fig. 3.2 that the maximum heat flux densities are as follows: in the region of the coolant inlet $q_{\text{пр}1} = 12.8$ kW/cm^2, in the region of the coolant outlet $q_{\text{пр}2} = 3.75$ kW/cm^2, with $W_1 = 0.1$–0.4 μm, and $W_2 = 0.4$–3.5 μm. In the region of the maximum heat flux density $W_2 = 1.5$–2 μm, which is in most cases much higher than the maximum allowable deformation of the reflectors for CO_2 lasers.

With the increase in coolant flow (Fig. 3.2b) W_2 can be dramatically reduced by using the materials with high bulk porosity (0.7–0.8) and the transition to a larger grain size. In this case, $q_{max} = 2$ kW/cm^2 at $W = 0.5$ μm. Powder-material structures with $\Pi_V > 0.75$ for the power optics are very difficult to produce. With a decrease in porosity to the real values (~0.6) the heat flux densities being removed will be ~1.4 kW/cm^2 at a 0.5 μm deformation. Note that increasing the coolant flow rate is also achieved through rational design of the coolant inlet and outlet system, providing a more uniform distribution of the coolant along the cooling surface.

One can see from Fig. 3.2c, d that the use of porous molybdenum allows an approximately four-fold decrease in the level of thermal deformations of mirror surfaces in both the first and second regions; in this case $q_{\text{пр}}$ somewhat decreases:

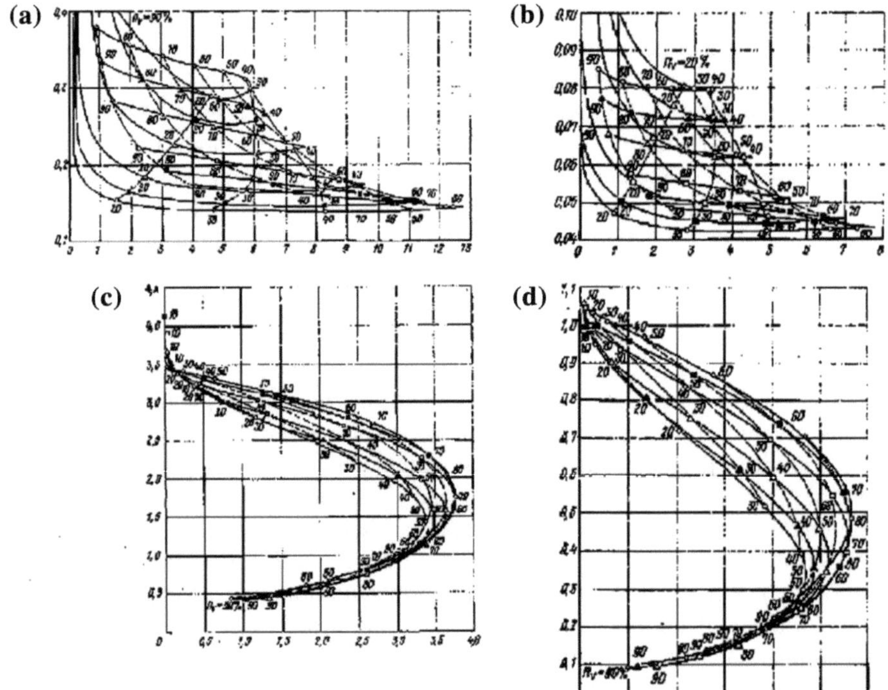

Fig. 3.2 Nomograms of thermal deformation characteristics of a family of water-cooled reflectors made of copper (**a**, **b**) and molybdenum (**c**, **d**) powders for the injection region (**a**, **c**) and the outflow region (**b**, **d**); points in this and other figures correspond to the following particle diameters or felt: $W - 20$, $W - 30$, $W - 50$, $W - 150$, and $W - 200$ μm

$q_{\text{пр}1} = 7.8$ kW/cm^2, ($q_{\text{пр}2} = 2.8$ kW/cm^2). For this structure (Fig. 3.2c) in the implementation of the maximum removed flows the arising thermal deformations are below the characteristic optical damage thresholds of the reflectors for CO_2 lasers. For example, a structure with $\Pi_V = 83$ % и $d_s = 20$ μm facilitates the removal of $q_{\max} = 2.8$ kW/cm^2 at $W = 0.45$ μm. With increasing the coolant flow $W_2 \rightarrow W_1$ (Fig. 3.2d), $q_{\max}$ decreases. In this region the level of heat flows to be removed can be increased by the use of materials with the larger size of the particles; for example, the molybdenum powder structure with $\Pi_V = 83$ % и $d_s = 20$ μm facilitates the removal of $q_{\max} = 2.1$ kW/cm^2 at $W = 0.23$ μm.

Figure 3.3 shows the thermal deformation characteristics of the reflectors with porous structures made of copper (Fig. 3.3a, b) and molybdenum (Fig. 3.3c, d)

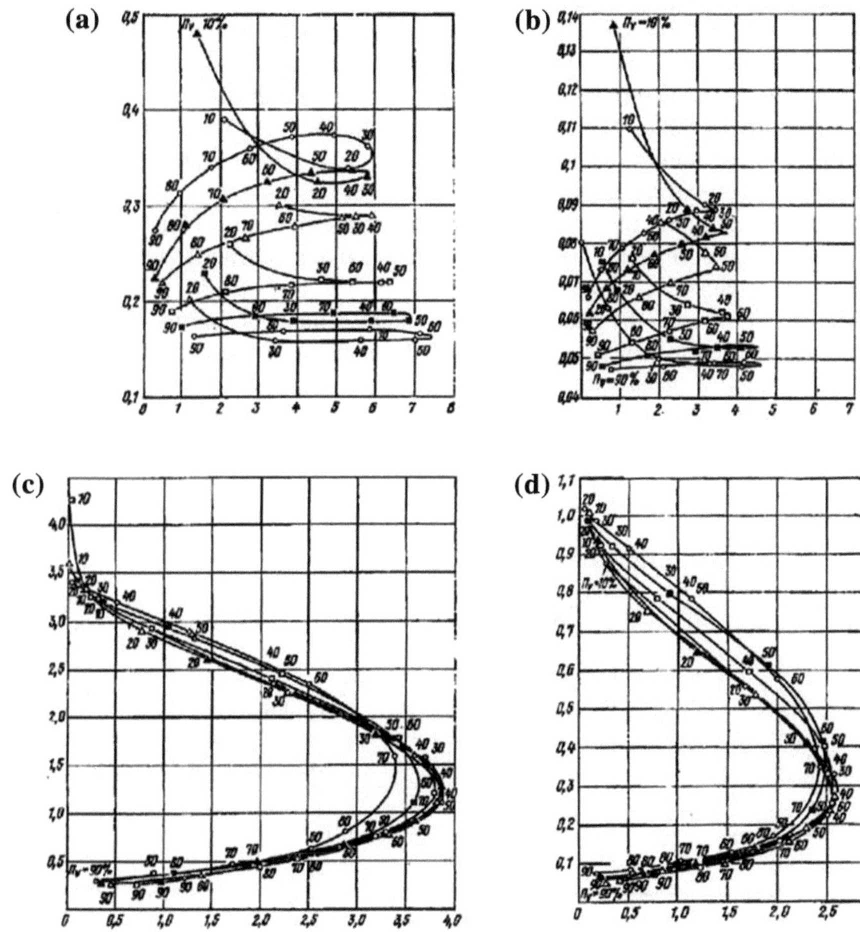

Fig. 3.3 Nomograms of thermal deformation characteristics of a family of water-cooled reflectors based on porous structures made of copper (**a**, **b**) and molybdenum (**c**, **d**) felt: the region of the coolant injection (**a**, **c**) and region of the coolant outflow (**b**, **d**)

felt. One can see from Fig. 3.3a, c that the transition to metal fibrous structures leads to a significant (~1.7) decrease in q_{pm}, which is associated with a decrease in the effective thermal conductivity of these structures compared to powder structures.

In regions of the coolant outflow (Fig. 3.3b, d) the maximum heat flux densities are on the same level as for the mirrors with powder structures; however, an increase in the coolant flow leads to a strong dependence of W on Π_V and to less sensitivity to the diameter of the initial fibers. Deformations corresponding q_{pm} are approximately two times lower compared to the powder structures: for copper and molybdenum reflectors $W = 1.2$ and 0.3 μm at $q_{pm} = 3.9$ kW/cm^2 and $q_{pm} = 2.6$ kW/cm^2, respectively.

For mirrors with a copper fiber structure the heat flux densities, which can be realized with $W = \lambda/20$ (where $\lambda = 10.6$ μm), of 2.4 kW/cm^2, are 2.4 kW/cm^2, which exceeds the corresponding level of q_{max} for powder materials. Thus, the use of metal fibrous structures for producing power optics elements is very promising; for example, the structure of molybdenum fibers with a diameter of 100–200 μm and a porosity of 50–60 % $q_{max} = 2.2$ kW/cm^2 at $W \approx 0.15$ μm. In addition, the possibility of obtaining materials with high Π_V (70–80 %) makes it possible to manufacture optimal reflectors.

We have compared the calculation results of thermal mirror-surface deformations of the laser reflector based on a porous copper-powder material with the experimental data.

Figure 3.4 presents the results of experimental studies of thermal deformations of a water-cooled mirror as a function of the density of the heat flow removed, and the results of calculation of the maximum thermal deformation characteristics in the region of the coolant removal for a family of reflectors with a 55–70 %

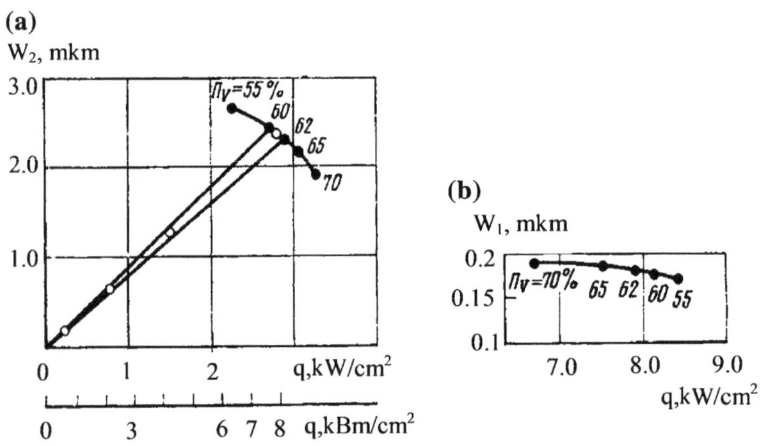

Fig. 3.4 Thermal deformation characteristics of a family of water-cooled reflectors based on porous structures made of copper powder: the region of the coolant injection (**a**), the region of the coolant outflow (**b**) (*filled circles* are the theory, *light circles* are the experiment)

porosity. Straight lines are the calculated characteristics of the tested reflector for the maximum porosity of 60 and 62 %. It is seen that there is good agreement between theoretical and experimental data [5–19].

Figure 3.4b shows the calculation of the thermal deformation characteristic of a family of reflectors in the region of the coolant inlet, which, like the curve in Fig. 3.4a, is obtained for the cooling surface temperature of the separating layer equal to 100 °C. It follows from Fig. 3.4b that $q_{\max} = 8.2\ \text{kW/cm}^2$, in good agreement with the experimental data.

Note in conclusion that this method of predicting the thermal and deformation characteristics of porous laser reflectors offers optimized type and parameters of the structure to ensure the necessary removal of heat fluxes at the admissible values of the distortion of the mirror surface. The results of theoretical and experimental studies suggest that water-cooled laser reflectors with porous structure have a high optical damage threshold. With the accumulation of new experimental data on convective heat transfer and hydrodynamics of flows in porous structures the latter can be quite easily introduced into the calculation algorithm.

3.7 Liquid-Metal Coolants in POEs Based on Porous Structures

In 1978, we were first to suggest that a further increase in the optical damage threshold of mirror surfaces of POEs based on porous structures is possible when liquid alkali metals and their alloys are used as coolants [7–11]. Prospects of utilising liquid-metal coolants in POEs were determined by the possibility of achieving a high heat transfer coefficient in the porous structure due to a favourable combination of thermophysical properties of liquid metals. This allowed one to lessen the requirements to the thermal conductivity of the porous structure material, which opened up the possibility of using new structural materials with a low thermal expansion coefficient and thermal conductivity in reflectors. Of particular interest was the employment of eutectic alloys of liquid metals with low melting points in POEs.

Consider some results of theoretical and experimental investigations of heat and thermal deformation characteristics of POEs cooled by the eutectic alloy Na–K. As part of earlier assumptions the heat transfer equation can be written as

$$\frac{d^2t}{dx^2} = \frac{h_e}{\bar{\lambda}} S_V (t - t_T), \tag{3.21}$$

where h_e is the heat transfer coefficient between the porous structure material and the coolant. Due to the lack of published data on the heat transfer of liquid metals in porous structures, the lower bounds of heat transfer coefficient were estimated

by using the known data on the heat transfer of liquid-metal coolants in triangular arrays of nuclear reactor fuel elements [15, 16]. To calculate the heat transfer of liquid metals in the nuclear fuel assemblies, use was made of the relations:
in densely packed structures ($s/d = 1$)

$$\mathrm{Nu} = \mathrm{Nu}_{\mathrm{lam}} + 0.0408\left(1 - 1/\sqrt{1.24\varepsilon + 1.15}\right) \mathrm{Pe}^{0.65}; \tag{3.22}$$

in not densely packed structures ($1.0 < s/d < 1.2$)

$$\mathrm{Nu} = \mathrm{Nu}_{\mathrm{lam}} + \frac{3.67}{90(s/d)^2}\left\{1 - \left[\frac{1}{\left[(s/d)^{30} - 1\right]/6 + \sqrt{1.24\varepsilon + 1.15}}\right] \mathrm{Pe}^{m_1}\right\}; \tag{3.23}$$

in not densely packed structures ($1.2 < s/d < 2$)

$$\mathrm{Nu} = \mathrm{Nu}_{\mathrm{lam}} + 3.67\mathrm{Pe}^{m_2}/90(s/d)^2. \tag{3.24}$$

Here, $m_1 = 0.56 + 0.19s/d - 0.1/(s/d)^{80}$; Pe is the Péclet number;

$$\mathrm{Nu}_{\mathrm{lam}} = \left[7.55\left(\frac{s}{d} - \frac{6.3}{(s/d)^{17(s/d)(s/d-0.81)}}\right)\right]\left[1 - \frac{3.6}{(s/d)^{20}(1 + 2.5\varepsilon^{0.86}) + 3.2}\right]$$

is the Nusselt number for the laminar flow; s/d is the relative spacing of the fuel elements in the array; and $\varepsilon = \lambda_{\mathrm{st}}/\lambda_{\mathrm{m}}$ is the ratio of the thermal conductivity of the fuel element cladding material to the thermal conductivity of the coolant. The relations (59)–(61) are valid for $\varepsilon > 0.01$ and $1 \leq \mathrm{Pe} \leq 4{,}000$.

Assuming that the hydraulic diameter of the array of the fuel elements corresponds to the hydraulic diameter of the POE porous structure ($d_s = d_p$), and the diameter of a set of rods—to the wire diameter (for metal-fibrous porous structures), we can obtain the dependence $d_s = d_m\Pi_V/(1 - \Pi_V)$ for felt porous structures.

Figures 3.5 and 3.6 show the results of numerical calculations of thermal deformation characteristics of the POE cooled by the eutectic coolant Na–K. It was assumed that the porous structures of the reflectors were made of molybdenum and invar felt. The mean diameter of the felt and the bulk porosity of the structure varied within $20 \leq d_m \leq 200$ μm and $0.1 \leq \Pi_V \leq 0.9$. The curves in Figs. 3.5 and 3.6 are the envelopes of the thermal deformation characteristics of the POE family and plotted at a constant pressure drop of the coolant and a maximum temperature of the cooling surface equal to 100 °C.

One can see from Fig. 3.2 that the deformation of the optical surface in the region of the coolant outlet, calculated with account for its heating in the porous structure, substantially exceeds the deformation in the region of the coolant inlet $\left(W_2^* > W_1^*\right)$. The maximum power densities of the heat flux for the POE in question are as follows: $q_1 > 20$ kW cm^{-2} in the region of the coolant inlet and $q_1 = 6.6$ kW cm^{-2}; in this case, $W_2^* = 0.3$ μ m. The minimum deformation W_2^* in the region of the coolant outflow at a power density of 4.2 kW cm^{-2} is 0.12 μm,

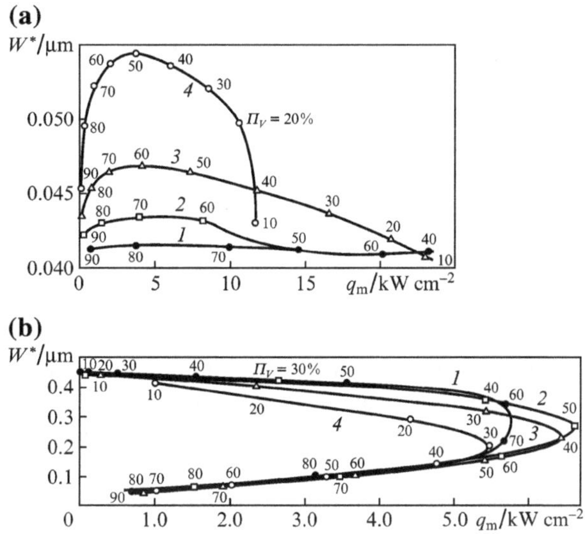

Fig. 3.5 Nomograms of thermal deformation characteristics of POEs based on metal-fibrous porous structures made of molybdenum, which are cooled by a Na–K coolant in the regions of its inlet (**a**) and outlet (**b**) at d_m = *1* 20, *2* 50, *3* 100 and *4* 200 μm

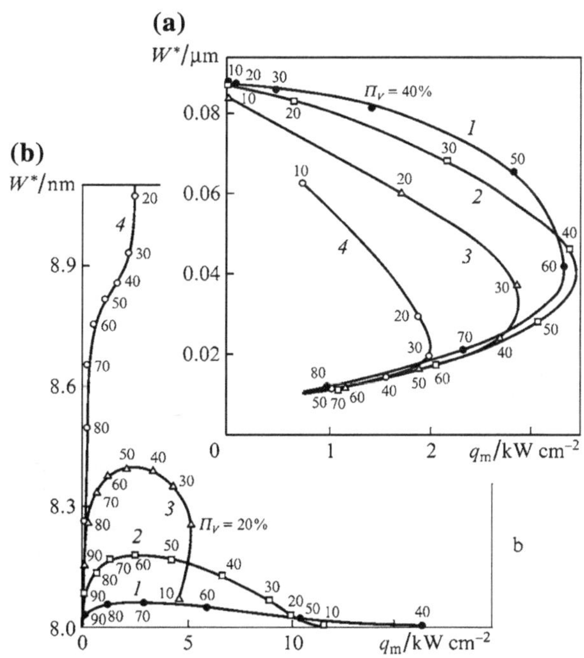

Fig. 3.6 Nomograms of thermal deformation characteristics of POEs based on metal-fibrous porous structures made of invar, which are cooled by a Na–K coolant in the regions of its inlet (**a**) and outlet (**b**) at d_m = *1* 20, *2* 50, *3* 100 and *4* 200 μm

which is significantly lower than the optical damage threshold of the POEs for CO_2 lasers.

Analysis of the data in Fig. 3.6 shows that the use of porous structures made of materials with a low thermal expansion coefficient (invar fibres) allows one to significantly (approximately by 3–4 times) reduce thermal deformations of the mirror surface both in the region of the coolant inlet and outlet in the case of liquid-metal cooling. Thus, the maximum thermal loads, experimentally allocated from the mirror surface, exceeded 10 kW cm^{-2}. The experimentally measured thermal deformations of POEs made of invar fibres in the region of minimum deformations were less than 0.5 μm.

It should be noted that the results presented in Figs. 3.5 and 3.6 clearly show that liquid metals are very promising for POE cooling. Such cooling in combination with porous structures made of materials with relatively low coefficients of thermal expansion opens up fundamentally new possibilities for creating a class of very precise POEs with a high optical damage threshold.

Today, due to the accumulation of experimental data on convective heat transfer and hydrodynamics in porous structures, such structures are widely used in space instrumentation and nuclear power systems exposed to high radiation doses. Due to the structural features, metal porous structures have no blind pores, which eliminates unwanted thermal processes. They provide good permeability, high thermophysical characteristics, ability to use POEs at a boiling point of working fluids in heated regions, high heat transfer rates and high limiting values of critical heat fluxes. Metal porous structures exhibit good physicomechanical and performance characteristics. Metallurgical production technology ensures their stability and reproducibility, long service life, and high reliability. One of the first mirrors based on porous structures [20–23] is shown in Fig. 3.7.

The new areas of research, which have been successfully developed recently, include the study of boiling on surfaces with porous coatings with their structural and hydrodynamic characteristics taken into account, the study of the influence of these characteristics on the contact thermal resistance between the porous and solid layers, and the study of heat transfer during condensation of liquids on the working surfaces of porous structures. It should be noted that our investigations of heat transfer in porous structures made it possible to develop the technological basis for creating

Fig. 3.7 Cooled POE with a powder porous structure

a series of water-cooled mirrors for power lasers by employing chemical etching of metal foils with subsequent soldering to fabricate a multilayer heat exchanger with a moderate degree of development of the heat exchange surface [16–56].

References

1. V.V. Apollonov, A.M. Prokhorov, V.Yu. Khomich, S.A. Chetkin, J. Sov, Quantum Electron. **11**, 1344 (1981)
2. V.V. Apollonov, P.I. Bystrov, V.F. Goncharov, A.M. Prokhorov, Patent No 135237
3. T. Yokobori, *An Interdisciplinary Approach to Fracture and Strength of Solids* (Mir, Moscow, 1971)
4. V.V. Apollonov, A.I. Barchukov, V.I. Borodin, P.I. Bystrov, V.F. Goncharov, A.M. Prokhorov et al., Sov. J. Quantum Electron. **8**, 672 (1978)
5. V.V. Apollonov, A.I. Barchukov, Y.A. Broval'skii, A.M. Prokhorov et al., Authentication Certificate No 135238 (priority date 08.12.1978)
6. V.V. Apollonov, A.I. Barchukov, P.I. Bystrov, V.F. Goncharov, A.M Prokhorov, Authentication Certificate No 144371 (priority date 19.01.1979)
7. V.V. Apollonov, P.I. Bystrov, V.F. Goncharov, A.M. Prokhorov et al., Authentication certificate no. 142696 (priority date 29.06.1979)
8. V.V. Apollonov, A.I. Barchukov, A.M. Prokhorov, in *Proceedings on Laser Optics Conference*, Leningrad, 1980
9. V.V. Apollonov, High power optics, Doctoral Dissertation, Moscow, FIAN, 1982
10. V.V. Apollonov, P.I. Bystrov, Y.A. Broval'skii, V.F. Goncharov, A.M. Prokhorov, Sov. J. Quantum Electron. **11**, 796 (1981)
11. V.V. Apollonov, P.I. Bystrov, Y.A. Broval'skii, V.F. Goncharov, A.M. Prokhorov, Sov. J. Quantum Electron. **11**, 798 (1981)
12. V.P. Isachenko, V.A. Osipov, A.S. Sukomel, *Heat Transfer* (Energiya, Moscow, 1969). (Mayorov V.A. Teploenergetika, (1), 64 (1978))
13. P. Forcheimer, Vereines deutscher Ingenieure, **45**, 1781–1788 (1901)
14. G.S. Beavers, E.M. Sparrow, J. Appl. Mech. **36**, 711 (1969)
15. S.V. Belov, *Porous Materials in Mechanical Engineering* (Mashinostroenie, Moscow, 1967)
16. V.I. Subbotin, M.Kh. Ibragimov, P.A. Ushakov, *Hydrodynamics and Heat Exchange in Nuclear Reactors* (Atomizdat, Moscow, 1975)
17. V.V. Apollonov, Laser Phys. **23**, 1 (2013)
18. W.M. Kays, A.L. London, *Compact Heat Exchangers* (McGraw Hill, New York, 1984; Energiya, Moscow, 1967)
19. J. Dupuit, *Etudes Theoretiques et Pratiques sur le Movement des Eaux* (Dunot, Paris, 1863)
20. V.V. Apollonov, B.F. Bunkin, L.N. Zakhar'ev, N.N. Polyashov, A.M. Prokhorov, Patent No 152944
21. V.A. Alekseev, V.N. Antciferov, V.V. Apollonov, A.M. Prokhorov, Patent no 1026594 (1983)
22. V.V. Apollonov, I.V. Goncharenko, N.E. Ivanov, E.V. Khristyan, Patent No 1046984 (1983)
23. V.V. Apollonov, A.I. Ivlutin, V.N. Motorin, A.M. Prokhorov, V.N. Kharchenko, C.A. Chetkin, A.A. Shubin, S.F. Sholev, Patent No 225192 (1984)
24. V.Y. Khomich, Candian Dissertation, Moscow, FIAN, 1981
25. V.V. Apollonov, V.I. Borodin, A.S. Brynskikh, S.V. Muravev, S.N. Temnov, S.A. Chetkin, Sov. J. Quantum Electron. **11**, 1344 (1981)
26. V.V. Apollonov, S.A. Chetkin, V.Y. Khomich, A.M. Prokhorov, Lett. JTP**7**, 1388 (1982)
27. Y.P. Voinov, Candian Dissertation, Almaz, 1982
28. S.A. Chetkin, Candian Dissertation, Moscow, GPI RAS, 1983
29. V.V. Apollonov, A.M. Prokhorov, A.G. Kostornov, L.M. Ostrovskaya, S.A. Chetkin, in *Proceedings of the Conference*, Kiev, USSR, 1984

30. V.V. Apollonov, V.I. Borodin, S.A. Chetkin, A.M. Prokhorov, V.V. Ostanin, Izvestiya of AS USSR. Ser. Physics **8**, 48 (1984)
31. V.V. Apollonov, S.A. Chetkin, V.Y. Khomich, A.M. Prokhorov, Izvestiya of AS USSR, Ser. Phys. **47**, 10, (1984)
32. V.V. Apollonov, S.A. Chetkin, V.Y. Khomich, A.M. Prokhorov, Boulder damage symposium XVI, NBS-Boulder, Colorado, USA, 1984
33. V.V. Apollonov, P.I. Bystrov, S.A. Chetkin, V.F. Goncharov, V.Y. Khomich, A.M. Prokhorov, Boulder damage symposium XVI, NBS-Boulder, Colorado, USA, 1984
34. V.V. Apollonov, P.I. Bystrov, S.A. Chetkin, V.F. Goncharov, V.Y. Khomich, A.M. Prokhorov, Boulder damage symposium XVI, NBS-Boulder, Colorado, USA, 1984
35. V.V. Apollonov, I.V. Goncharenko, N.E. Ivanov, E.V. Khristyan, И., Izvestiya of AS USSR. Ser. Phys. **48**, 8 (1984)
36. V.A. Alekseev, V.N. Antciferov, V.V. Apollonov, A.M. Prokhorov, S.V. Bilibin, A.P. Kunevich et al., Lett. JTP, **11**, B. 22 (1985)
37. V.V. Apollonov, Science and Life, No 9,50, 1985
38. V.V. Apollonov, G.I. Babayants, Y.P. Voinov, A.M. Prokhorov, E.V. Khristyan, B.S. Kishmakhov, Patent No 1277854 (1985)
39. V.V. Apollonov, G.I. Babayants, P.I. Bystrov, V.F. Goncharov, A.M. Prokhorov, Patent No 244286 (1985)
40. V.V. Apollonov, A.I. Ivlutin, V.N. Motorin, A.M. Prokhorov, V.N. Kharchenko, C.A. Chetkin, Patent No 1278565 (1985)
41. V.V. Apollonov, I.V. Goncharenko, A.M. Prokhorov, C.A Chetkin, E.V. Khristyan, A.A. Shubin, Patent No 1398717 (1985)
42. V.V. Apollonov, I.V. Goncharenko, A.M. Prokhorov, C.A. Chetkin, E.V. Khristyan, A.A. Shubin, Patent No 1301260 (1986)
43. V.V. Apollonov, A.S. Brynskih, S.A. Chetkin, A.M. Prokhorov, S.N. Temnov, Patent No 240743 (1986)
44. V.V. Apollonov, V.I. Borodin, A.S. Brynskih, S.A. Chetkin, A.M. Prokhorov, S.N. Temnov, Patent No 234217 (1986)
45. V.V. Apollonov, V.I. Borodin, A.S. Brynskih, S.A. Chetkin, V.V. Ostanin, A.M. Prokhorov, A.M. Temnov, Mirror, Patent No 234216 (1986)
46. V.V. Apollonov, V.A. Alekseev, S.V. Bilibin, Y.V. Danchenko, S.A. Chetkin, A.P. Kunevich, A.M. Prokhorov, Mirror, Patent No 241100 (1986)
47. V.N. Antciferov, V.V. Apollonov, V.A. Alekseev, S.V. Bilibin, Y.V. Danchenko, S.A. Chetkin, A. P. Kunevich, A.M. Prokhorov, Patent No 261206 (1986)
48. V.V. Apollonov, A.S. Brynskii, M.S. Granovskii, L.M. Ostrovskaya, A.M. Prokhorov, C.A. Chetkin, V.N. Rodin, M.I. Tsipin, A.A. Shubin, S.F. Sholev, Patent No 251080 (1987)
49. V.V. Apollonov, M.S. Granovskii, N.N. Zubkov, V.N. Motorin, L.M. Ostrovskaya, V.N. Rodin, A.M. Prokhorov, V.N. Kharchenko, C.A. Chetkin, Patent No 1455217 (1987)
50. V.V. Apollonov, B.F. Belov, V.N. Karmatskii, L.M. Ostrovskaya, A.M. Prokhorov, C.A. Chetkin, V.N. Rodin, M.I. Tsipin, S.F. Shevakin, S.F. Sholev, Patent No202719 (1987)
51. V.V. Apollonov, M.S. Granovskii, N.N. Zubkov, V.N. Motorin, A.M. Prokhorov, V.N. Kharchenko, C.A. Chetkin, Patent No 4198992 (1987)
52. V.V. Apollonov, V.I. Andryushin, N.N. Zubkov, V.N. Motorin, A.M. Prokhorov, V.N. Kharchenko, C.A. Chetkin, A.A. Shubin, V.A. Shurygin, Patent, No 4249488 (1987)
53. V.V. Apollonov, A.I. Ivlutin, V.N. Motorin, A.M. Prokhorov, V.N. Kharchenko, C.A. Chetkin, Patent, No 4212816 (1987)
54. V.V. Apollonov, A.I. Ivlutin, V.N. Motorin, A.M. Prokhorov, V.N. Kharchenko, C.A. Chetkin, Patent, No 4227164 (1987)
55. V.V. Apollonov, V. I. Andryushin, N.N. Zubkov, V.N. Motorin, C.A. Prokhorov Chetkin, A.A. Shubin, V.A. Shurygin, Patent No 288490 (1987)
56. V.V. Apollonov, Y.V. Danchenko, S.A. Chetkin, A.M. Prokhorov, Patent No 261206 (1987)

Chapter 4
Adaptive POEs and Optical Systems Based on Them

The variety of the phenomena that change the optical characteristics of the medium in the propagation path of radiation and in the optical system of the laser leads to degradation of the quality of the wavefront (WF), which is manifested by a significant increase in the angular divergence of the generated beam and by a reduction in the peak intensity upon focusing. Most fully the entire range of requirements to WF correctors in adaptive optics systems is met by POEs with adjustable shape of the reflecting surface, in which the WF distortion are compensated for by changing the shape of the mirror surface. In this case it is possible: (1) to fabricate cooled and uncooled adaptive POEs with a high optical damage threshold in a wide range of radiation exposures [1–58]; (2) to produce adaptive optics systems for the entire set of currently known schemes for generation of cw and repetitively pulsed laser radiation in wavelengths ranging from far-IR to ultraviolet; and (3) to manufacture adaptive POEs for correcting and measuring nonstationary phase distortions in the time interval up to several milliseconds, by selecting the substrate materials of the mirror surface which provide their predetermined static and dynamic deformation properties.

The most challenging, in our view, is the realisation of adaptive POEs with a high optical damage threshold of the reflecting surface, because it is necessary in this case to combine the shape and cooling control systems in the reflector. Our approach to creating adaptive POEs was based on the methods of forced heat removal for cooling the mirror surface while shaping the reflecting surface by controlled elastic deformation of the porous structure of the heat exchanger.

Prospects of our proposal consisted in the possibility of providing the necessary static and dynamic deformation and thermal characteristics of adaptive POEs, because the use of porous structures allows one to implement the optical damage threshold (up to several tens of kW cm^{-2}) of the mirror surface, whereas their use as a substrate material of the mirror surface having low stiffness enables control of the shape of the reflecting surface in a wide range of local displacement amplitudes of its individual regions. Moreover, since in the operating condition the material of the porous heat exchanger of an adaptive POE is filled with a liquid coolant, natural resonant oscillations of the mechanical design of the adaptive POE may be effectively damped in the dynamic regime of the device operation.

V.V. Apollonov, *High-Power Optics*, Springer Series in Optical Sciences 192,
DOI 10.1007/978-3-319-10753-0_4

Studies on modelling the correction of basic WF distortions by the adaptive POE, determining the optical quality of the intense laser radiation flux (including the WF tilt, defocusing, spherical aberrations), showed that at a consistent satisfaction of the power optics requirements, involving realisation of high values of the optical damage threshold (the porous structure thickness of the exchanger must be several millimetres), inaccuracy of conjugation of the WF shape with the shape of the reflecting surface for CO_2 laser radiation is $\lambda_0/10 - \lambda_0/20$ using a control system with 50–60 actuators on the aperture of the adaptive POE up to 100 mm in length.

Along with the well-known solutions, our approach to creating adaptive POEs is very promising in the development of adaptive optics systems for high-power lasers [59–72]. However, its implementation required complex investigations to establish the peculiarities of dynamic and static regimes of deformation of porous structures, to study the influence of the processes of internal friction in porous materials on the dynamics of their cyclic loading, to determine the effect of anisotropy of mechanical properties of the structure on the form of the response function of the reflecting surface, to establish an optimal (for these devices) control of an adaptive optical system, and to create new types of actuators with high energy capacity. It is important to note the major role of the design bureaus "Almaz" and "Astrophysics" headed at that time by B.V. Bunkin and N.D. Ustinov in achieving these goals.

Great importance in the development and creation of adaptive POEs with specified static and dynamic characteristics was given to actuators providing the required amplitudes of deformation over a wide dynamic range. Solutions related to the use of piezoelectric materials in physical problems associated with adaptive POEs are not free from drawbacks. These include the need for high strains required for the realisation of amplitude displacements and the inevitability of hysteresis phenomena that hinder the formation of a phase-conjugated laser beam WV by the relief of the reflecting surface. In the regimes of 'modulation' and 'phase conjugation' the amplitudes of local displacements of the reflecting surface should reach $0.1 - 0.5\lambda_0$ and $1 - 5\lambda_0$, respectively. To ensure such displacement amplitudes we proposed adaptive POE actuators of new types, which are made of magnetostrictive materials and implement the conditions for the Joule and Wiedemann effects [73–78]. At the same time we pointed out the prospect of creation of compact highly efficient actuators, providing a stable amplitude displacement in the frequency range up to 10 kHz.

The employment of the designed and built adaptive POEs utilising porous structures is not confined to adaptive optics, although in this field they solve a number of important problems. According to the results of modelling intracavity optical systems [78–86], the use of adaptive POEs allows one to obtain the diffraction angular divergence of radiation fluxes when use is made of unstable resonators in high-power carbon dioxide laser systems. Adaptive POEs were essentially a new type of devices ensuring the local control of phase characteristics of coherent radiation fluxes. As a result, they served as prototypes for different laser beam modulation, selection and scanning devices. For example, the use of an adaptive POE in the laser cavity made it possible to convert high-power CW radiation into high-frequency repetitively pulsed radiation by *Q*-switching [87–105], and the employment of adaptive POEs in a Fabry–Perot interferometer allowed for automated analysis of spectral and modal composition of laser radiation, etc. Undoubted are the advantages of this class of adaptive POEs in traditional applications of adaptive systems, such as laser ranging. Here we should mention paper [106, 107], which presents the characteristics of a number of adaptive mirrors.

4.1 POEs Based on Porous Structures

Generation and transmission of laser radiation with a near-diffraction-limited divergence is one of the urgent problems of modern power optics. In this part of the review we present the results of the studies on the creation of phase correctors of the beam path for laser systems based on POEs being cooled [79–86]. The shape of the reflecting surface is controlled in this case by the deformation of the porous structure of the heat exchanger. The field of application of phase correctors based on cooled POEs is not confined to adaptive optics. For example, the use of adaptive POEs in optical systems of laser ranging can significantly extend the capabilities of the latter due to the generation of the beams with predetermined optimal wavefront properties.

The feasibility of controlling the shape of the reflecting surface of the cooled POE on the basis of a porous structure was considered under the following model assumptions: a porous heat exchanger of radius R and thickness H ($H \ll R$) is rigidly connected to the base and the side surface of the body, which is made of an absolutely rigid (in comparison with the porous structure) material, except for those portions to which the system of actuators in the form of cylinders of radius $a \ll R$ is connected. The actuators are also made of an absolutely rigid (in comparison with the porous structure) material capable of producing the specified movements or stresses. The free surface of the porous structure has a separation layer with a deposited reflective coating [92–100,108, 109–116].

The assumption about the absolute rigidity of the POE body and actuators allows one to neglect the deformations of the latter during the formation of a given relief of the reflecting surface $W_{\text{refl}}(r,\varphi)$, defined as follows: for the incident beam with the wavefront $\Phi_1(r,\varphi)$, the reflecting surface is formed in such a way that reflected beam has the wavefront $\Phi_2(r,\varphi)$, i.e.,

$$W_{\text{refl}}(r,\varphi) = \frac{1}{2}[\Phi_2(r,\varphi) - \Phi_1(r,\varphi) + N\lambda], \quad (4.1)$$

where λ is the laser wavelength and N is an integer.

Without loss of generality, the main types of $\Phi_1(r,\varphi)$ include the slope of the wavefront, defocusing and spherical aberration, and as a function of a predetermined shape of the wavefront of the reflected beam use is made of the plane, i.e., $\Phi_2(r,\varphi) = 0$.

For given Φ_1 and Φ_2, the formation of the reflecting POE surface profile W_{refl} by the control system, containing a finite number of actuators, is not possible. It is therefore necessary to find the optimal arrangement of the movable structure of actuators, namely, their number and mutual position in order to ensure the required accuracy of conjugation of the shape of the reflecting surface of the adaptive POE with basic types of wavefront aberrations of the laser beam.

In solving the formulated problem it was necessary to consider the deformation characteristics of individual actuators of the adaptive POE. Within the approximate linear theory of elastic deformation of the porous structure of the heat exchanger, the problem of determining the shape of the reflecting surface W_{refl}, arising from action the actuator located on the non-reflective side of the porous structure at a distance b from the axis of the mirror, is reduced to the equation

$$\Delta_{\perp} W + \frac{2(1-\nu)}{1-2\nu}\frac{\partial^2 W}{\partial z^2} = 0, \tag{4.2}$$

$$W\Big|_{z=0} = U_0 \Theta(a^2 - (b-x)^2 - y^2), \quad \frac{\partial W}{\partial z}\Big|_{z=H} = 0, \quad W|_{r=R} = 0$$

(where ν is Poisson's ratio, Θ is the Heaviside function, u_0 is the value of the actuator displacement), whose solution for $W_{\text{refl}} = W(z = H)$ has the form

$$W_{\text{refl}} = 8u_0 \frac{a}{R} \sum_{k=0}^{\infty} \sum_{j=1}^{\infty} A_{kj} \frac{J_k\left(\mu_j^{(k)} \frac{r}{R}\right) \cos k\varphi}{\mu_j^{(k)} \operatorname{ch} D_{jk} (1 + \delta_{k,0})}, \tag{4.3}$$

where $A_{kj} = \frac{J_k\left(\mu_j^{(k)} \frac{b}{R}\right) J_1\left(\mu_j^{(k)} \frac{a}{R}\right)}{J_{k+1}^2(\mu_j^{(k)})}$, $J_k\left(\mu_j^{(k)} \frac{b}{R}\right)$ is the Bessel function of the kth order, $\mu_j^{(k)}$ is the jth root of the equation $J_k(\mu_j^{(k)}) = 0$, and $D_{jk} = \mu_j^{(k)} \sqrt{\frac{1-2\nu}{2(1-\nu)}} \frac{H}{R}$, and the force which must be applied to the actuator is defined as

$$P = \iint\limits_{S':(x-b)^2+y^2 \le a^2} \sigma_{zz}\Big|_{z=0} dS' = -8\pi u_0 R E \sqrt{\frac{2(1-\nu)}{1-2\nu}} \left(\frac{a}{R}\right)^2 \frac{1}{1+\nu}$$
$$\times \sum_{k=0}^{\infty} \sum_{j=1}^{\infty} A_{kj}^2 \operatorname{th} D_{jk} / (\mu_j^{(k)} (1 + \delta_{k,0})), \tag{4.4}$$

where E is Young's modulus of the porous structure of the heat exchanger.

Analysis of the presented dependences shows that the transfer deformation function (i.e., the response function of the reflecting POE surface to the action of the actuator) has a smooth Gaussian-like shape. The width of the response function is comparable with the actuator radius (Fig. 4.1).

The dependence of the response function maximum on the parameters $AR = a/R$ and $HR = H/R$ is shown in Fig. 4.2, and the dependence of the response function width Δ at the level of 1/2 is given in Fig. 4.3. The dependence of the dimensionless specific pressure of form

$$\sigma = \sqrt{\frac{1-2\nu}{2(1-\nu)}} \frac{PR}{16\pi G u_0 a^2},$$

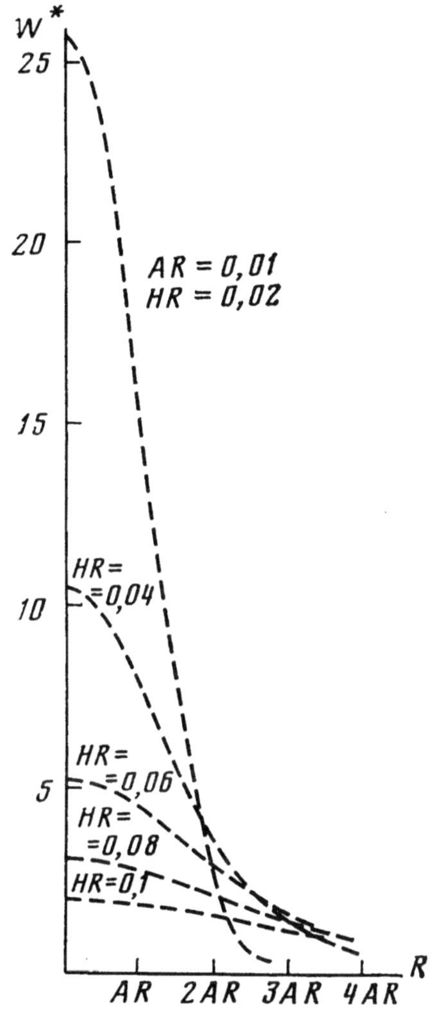

Fig. 4.1 The response function of reflecting surface on single actuator action as a function of *AR* and *HR*

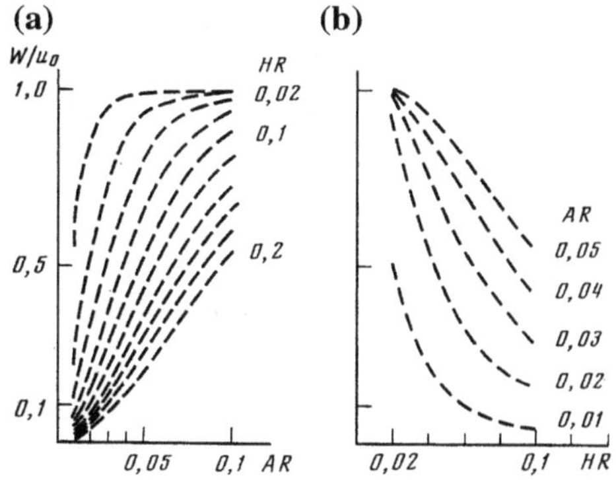

Fig. 4.2 The dependence of the response function maximum on the parameters $AR = a/R$ and $HR = H/R$

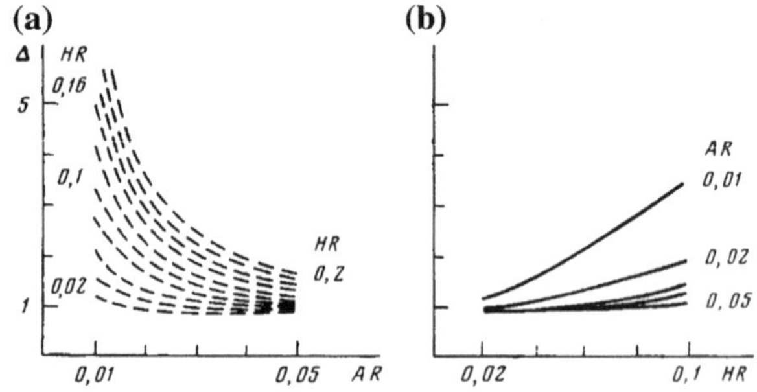

Fig. 4.3 The dependence of the response function width Δ at the level of ½

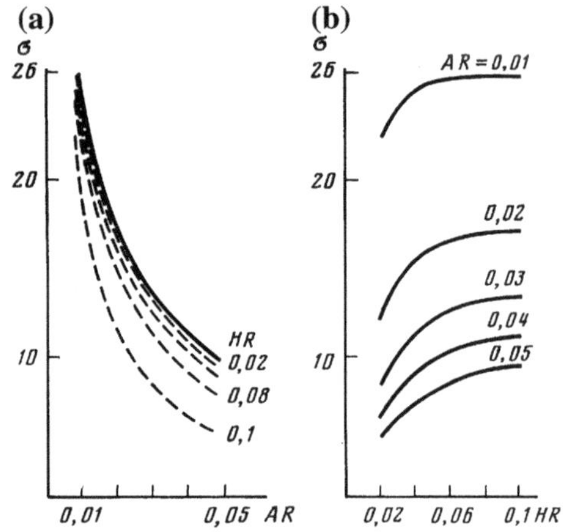

Fig. 4.4 The shear modulus of the porous structure, on the parameters *AR* and *HR*

where *G* is the shear modulus of the porous structure, on the parameters *AR* and *HR* is shown in Fig. 4.4.

From the figures it follows that in the case of broad response functions with $\Delta > 1$, its maximum value is $(W_0/u_0) < 1$, whereas in the case of narrow response functions of the reflecting surface ($\Delta = 1$), W_0/u_0 reaches its maximum value equal to 1. Herein, the value of σ tends to its minimum. According to the dependences in Fig. 4.2a, b, the maximum of W_0/u_0 is reached at the *AR* values, such that $AR \geq 1.5\,HR$.

For adaptive systems operating in real time, the reflecting POE surface should vary according to the change in the phase distortions along the laser beam propagation path, whose times of the 'frozen' existence constitute only 0.1–1 ms. Therefore, the width of the frequency range of regulated impacts of actuators of the adaptive POE must be 1–10 kHz and in the deformable porous structure should not show any resonance effects.

We determined the spectrum of natural oscillations of the porous structure by using an approximate dynamic equation of elasticity:

$$\frac{1}{c^2}\frac{\partial^2 W}{\partial t^2} = \Delta_{\perp} W + \frac{2(1-\nu)}{1-2\nu}\frac{\partial^2 W}{\partial z^2}, \tag{4.5}$$

on the basis of which the eigenfrequency spectrum of the adaptive POE is defined as

$$\omega_{mjl} = \frac{c}{R}\left[\mu_j^{(m)2} + \frac{\pi^2}{4}(1+2l)^2\left(\frac{R}{H}\right)^2\frac{2(1-\nu)}{1-2\nu}\right]^{1/2} > \pi\frac{c}{H}, \tag{4.6}$$

where c is the speed of sound in the porous material.

The position of nodes and antinodes on the nonreflecting side of the porous structure is found from the expressions

$$\begin{aligned} J_m\left(\mu_j^{(m)}\frac{r_{mjl}}{R}\right) &= 0, \quad m\varphi_{mjl}^{y^3} = \frac{\pi}{2}(1+2l), \\ J_m{}'\left(\mu_j^{(m)}\frac{r_{mjl}^{\pi\text{yq}}}{R}\right) &= 0, \quad m\varphi_{mjl}^{\pi\text{yq}} = \pi l \end{aligned} \tag{4.7}$$

Thus, if the range of operating frequencies of the actuators does not reach $\pi\frac{c}{H}$, the resonant effects which influence the stability of the control system of the adaptive POE will be not observed. If the operating frequencies exceed the value of $\pi\frac{c}{H}$, then one way to address the resonance effects is to place the actuators in the nodes of the corresponding natural oscillations of the porous structure.

The analysis of the stress state shows that the shape of the reflecting surface of the POE is a Fourier series over the system of functions $\left\{J_k\left(\mu_j^{(k)}\frac{r}{R}\right)\cos k\varphi\right\}$ on a circular aperture of radius R, i.e., any phase distortion represented by the function $F(r,\varphi)$, such that the integral $\iint\limits_S F^2(r,\varphi)dS$ exists, can be adjusted to any degree of precision ε, using an appropriately designed system of actuators, where ε is defined by the number of the actuators and their relative position.

In the process of adjustment, the shape of the reflecting surface has the form:

$$W^*_{\text{refl}} = \sum_{n=1}^{N} u_n\Psi_n(b_n,\varphi_n,r,\varphi), \tag{4.8}$$

where $\{b_n,\varphi_n\}$ are the coordinates of the center of the nth actuator on the nonreflective side of the porous structure, (r,φ) are the coordinates of the point of the reflecting surface, and $u_n\Psi_n$ is the response function of the reflecting surface to the action of the nth actuator. The problem of optimal conjugation of the shape of a reflecting POE surface with a flux having a wavefront $F(r,\varphi)$ is reduced to the determination of the uп and $\{b_n,\varphi_n\}$, for which the conjugation error, defined as

$$\varepsilon = \frac{1}{S}\iint\limits_S\left[F(r,\varphi) - \sum_{n=1}^{N} u_n\Psi_n(b_n,\varphi_n,r,\varphi)\right]^2 W_1(r\varphi)dS \tag{4.9}$$

[where N is the number of actuators, and $W_1(r,\varphi)$ is the aperture function defining the region of the reflecting POE surface which is adaptively controlled], would be minimal.

According to the method of least squares, the minimum ε is achieved on a set of values $\{u_n\}$, if the following conditions $(\partial\varepsilon/\partial u_n) = 0$ are fulfilled for any $1 \le n \le N$.

The vector $\{u_n\}$, minimizing the functional ε, is a solution to a system of linear equations

$$\sum_{k=1}^{N} A_{kn}^{*} u_k = B_n, \tag{4.10}$$

where the elements of the matrix A_{kn}^{*} having the dimension $N \times N$ are defined as

$$A_{kn}^{*} = \frac{1}{S} \iint_{S} W_1 \Psi_k \Psi_n dS, \tag{4.11}$$

i.e., they are characteristic of the number and mutual arrangement of the system of POE actuators, and the elements B_n of the vector B having the dimension N

$$B_n = \frac{1}{S} \iint_{S} W_1 F \Psi_n dS \tag{4.12}$$

are determined not only by the POE design, but also by the form of the function describing the perturbation of the wavefront of the laser beam. The compensation error ε is defined as

$$\varepsilon = |F| - \sum_{n=1}^{N} u_n B_n, \tag{4.13}$$

where

$$|F| = \frac{1}{S} \iint_{S} W_1 F^2 dS.$$

We have defined the form of the elements $A_{kn}{}^{*}$, B_n, $|F|$ for the main types of aberrations. If the aperture function is $W_1(r, \varphi) = \Theta(R_1 - r)$, where R_1 is the radius of the 'active' sub-aperture and $R_1 < R$, the transfer matrix elements of the POE have the form

$$\begin{aligned} A_{nm}^{*} = & \sum_{k=0}^{\infty} \sum_{\substack{i,j=1 \\ i \neq j}}^{\infty} \frac{A_{kj}(b = b_n) A_{ki}(b = b_m) \cos k(\varphi_n - \varphi_m)}{\mu_j^{(k)} \mu_i^{(k)} \mathrm{ch}\, D_{ik} \mathrm{ch}\, D_{jk}} \\ & \times \frac{\mu_i^{(k)} J_k(\mu_j^{(k)} \gamma) J_{k-1}(\mu_i^{(k)} \gamma) - \mu_j^{(k)} J_k(\mu_i^{(k)} \gamma) J_{k-1}(\mu_j^{(k)} \gamma)}{\mu_j^{(k)2} - \mu_i^{(k)2}} \\ & + \frac{\gamma}{2} \sum_{k=0}^{\infty} \sum_{j=1}^{\infty} \frac{A_{kj}(b = b_n) A_{kj}(b = b_m) \cos k(\varphi_n - \varphi_m)}{\mu_j^{(k)2} \mathrm{ch}^2 D_{jk} (1 + \delta_{k,0})} \\ & \times (J_k^2(\mu_j^{(k)} \gamma) - J_{k+1}(\mu_j^{(k)} \gamma) J_{k-1}(\mu_j^{(k)} \gamma)), \end{aligned} \tag{4.14}$$

where $\gamma = R_1/R$.

If $\gamma \to 1$, i.e., the control is performed over the entire aperture of the adaptive POE:

$$A^*_{nm} = \sum_{k=0}^{\infty}\sum_{j=1}^{\infty} A_{kj}(b=b_n)\frac{J_k\left(\mu_j^{(k)}\frac{b_m}{R}\right)(\delta_{k,0}+\cos k(\varphi_n-\mu_m))}{\mu_j^{(k)2}(1+\delta_{k,0})^2 \mathrm{ch}\, D_{jk}}. \quad (4.15)$$

In the case of correction of the wavefront tilt, described by the expression

$$F(r,\varphi) = r\,\mathrm{tg}\,\alpha\, \sin\varphi\Theta(R_1-r), \quad (4.16)$$

where α is the angle of inclination of the wavefront plane, the element of the vector B_n has the form

$$B_n = -\mathrm{tg}\,\alpha\frac{RR_1}{8a}\sum_{j=1}^{\infty} A_{ij}(b=b_n)\frac{J_2(\mu_j^{(1)}\gamma)\sin\varphi_n}{\mu_j^{(1)2}\mathrm{ch}\, D_{j1}}, \quad (4.17)$$

and

$$|F| = 0,25\,\mathrm{tg}^2\,\alpha R^2\gamma^4. \quad (4.18)$$

When correcting a cylindrical deflection of a plane wavefront on the same subaperture $R_1 < R$, described by the function

$$F(r,\varphi) = \pm\frac{r^2\sin^2\varphi - R_1^2}{2R_2}\Theta(R_1-r) \quad (4.19)$$

(where R_2 is the radius of the cylindrical surface), the components of B_n are defined as

$$B_n = \frac{\gamma R^3}{32R_2 a}\left\{\sum_{j=1}^{\infty}\frac{A_{0j}(b=b_n)}{\mu_j^{(0)3}\mathrm{ch}\, D_{j0}}\left[\frac{4J_1(\mu_j^{(0)}\gamma)}{\mu_j^{(0)2}} - 2\gamma\frac{J_0(\mu_j^{(0)}\gamma)}{\mu_j^{(0)}}\right] + \frac{\gamma^2\cos 2\varphi_n}{2}\sum_{j=1}^{\infty}\frac{A_{2j}(b=b_n)}{\mu_j^{(2)3}\mathrm{ch}\, D_{j2}}J_3(\mu_j^{(2)}\gamma)\right\}, \quad (4.20)$$

$$|F| = 5R_1^6/(32R^2R_2^2). \quad (4.21)$$

When correcting the defocusing $F(r,\varphi) = \pm\frac{R^2-r^2}{2R_2}$, where R_2 is the radius of the spherical surface, the components have the form

$$B_n = \pm\frac{z\mathrm{max}}{a/R}\sum_{j=1}^{\infty}\frac{J_0\left(\mu_j^{(0)}\frac{b_n}{R}\right)J_1\left(\mu_j^{(0)}\frac{a}{R}\right)}{\mu_j^{(0)4}J_1(\mu_j^{(0)})\mathrm{ch}\, D_{j0}}. \quad (4.22)$$

where $z_{max} = R^2/(2R_2)$ and $|F| = R^4/(12R_2^2)$. Spherical aberration is defined as $F(r,\varphi) = \pm\frac{B}{4}\left(\frac{r^4}{R^4} - 1\right)$; therefore, the components have the form

$$B_n = \mp\frac{4z_{\text{MaKC}}}{a/R}\sum_{j=1}^{\infty}\frac{J_0\left(\mu_j^{(0)}\frac{b_n}{R}\right)J_1\left(\mu_j^{(0)}\frac{a}{R}\right)(\mu_j^{(0)2} - 4)}{\mu_j^{(0)6}J_1(\mu_j^{(0)})\text{ch}\,D_{j0}} \tag{4.23}$$

and $z_{max} = B/4$ and $|F| = 8z_{max}^2/15$. In the case of small-scale phase distortion corrections, which we modeled by a Gaussian function $F(r,\varphi) = u_1\exp\left(-K_0r^2\right)$, where $K_0 = 2/a_1^2$ and a_1 is the transverse size, defined at the level of $1/e^2$, such that $a_1 \leq a$ and $|F| = u_1a_1^2/(4R^2)$. The vector B is defined as

$$B_n = \frac{u_1a_1^2}{16aR}\sum_{j=1}^{\infty}\frac{A_{0j}(b = b_n)}{\mu_j^{(0)}\text{ch}\,D_{j0}}\exp\left(-\frac{\mu_j^{(0)2}a_1^2}{8R^2}\right). \tag{4.24}$$

From the above it follows that we constructed a model of an adaptive cooled POE based on a porous structure, solved the problem of conjugation of the shape of the reflecting surface with the main types of aberrations and derived the basic expressions that describe the algorithm of optimal phase conjugation. We developed a new class of adaptive POEs based on porous structures, which have an optical efficiency threshold typical of a mirror surface (with respect to the thermal flux) at the level of a few kW cm^{-2}, allowing one to provide the correction of the laser beam wavefront distortions over a wide time range with high spatial resolution. We proposed new types of actuators for their use in adaptive POEs based on magnetostrictive materials, in which the Joule and Wiedemann conditions are met [117, 118].

4.2 Formation of the Adaptive POE Surface Conjugate with Wavefront Distortions

Synthesis of the aberration-conjugated wavefront of a reflecting surface by controlled elastic deformation of the substrate at a finite number of points allows one to adjust the phase distortions of coherent radiation fluxes. This method is simple to implement and is widely used when creating precise adaptive POEs for high-power lasers and astronomical telescopes.

At present, adaptive POEs of mirror type were created with an aperture diameter of one meter and the number of control channels (i.e., the number of actuators) up to several hundreds [75–85]. The study of adaptive POEs showed the versatility and high efficiency of this method for obtaining aberration-free fluxes of coherent radiation. However, the problem of the accuracy of the reflecting surface of the adaptive POE, conjugated with various wavefront aberrations, has no solutions satisfying the practical requirements.

Any phase conjugated surface of the adaptive POE can be represented as a superposition of its response functions. Consequently, it was necessary to establish physically the models of the response function formation under the action of a

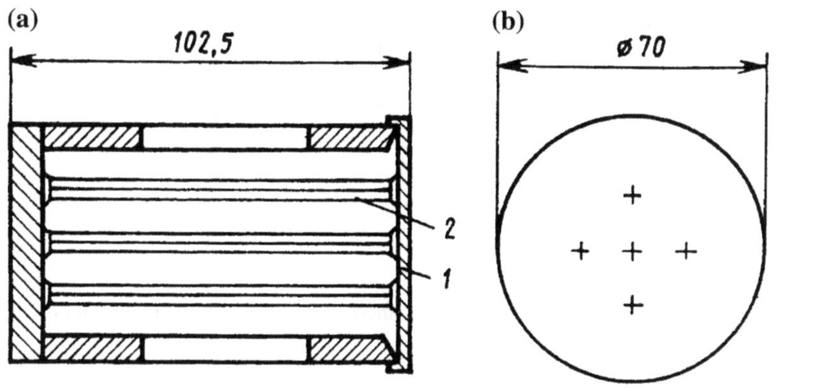

Fig. 4.5 Adaptive POE construction and piezoceramic based actuators distribution: *1*—reflecting surface, *2*—piezoceramics actuator

single actuator, adequately describing the characteristics of the deformation of the adaptive POE substrate by a strictional actuator and permitting an exact analytical interpretation.

We experimentally and theoretically investigated the response functions of the POEs in a model of a thin plate sealing the perimeter and resting in a finite number of points on elastic actuator supports; we also proposed a physical model which involves the numerical simulation of the formation of surface conjugated with wavefront aberrations.

In the experimental study we used the design of the adaptive POE with five actuators (Figs. 4.4 and 4.5) made of polarized piezoceramics PCR-6. Below we present the piezomechanical properties of PCR-6 and the actuator (Table 4.1).

Table 4.1 Piezomechanical properties of PCR-6 and the actuators based on that material

Relative permittivity	
$\varepsilon_{33}/\varepsilon_0$	2,300
Dielectric loss tangent tgδ	
At $E = 5$ kV/m	0.004
At $E = 100$ kV/m	0.01
Piezoelectric constant d_{31}, pC/N	195
Curie temperature T_c, °C	230
Young's modulus Y_{11}^E, GPa	75
Mechanical strength, $\sigma_{\text{раст}}$, MPa	50
Density ρ, t/m^3	8
Speed of sound V_1^E, km/s	3.02
Maximum applied voltage, V	±750
Electric capacity, nF	15 ± 10 %
Sensitivity (without external force), nm/V	±25
Maximum external force, N	60

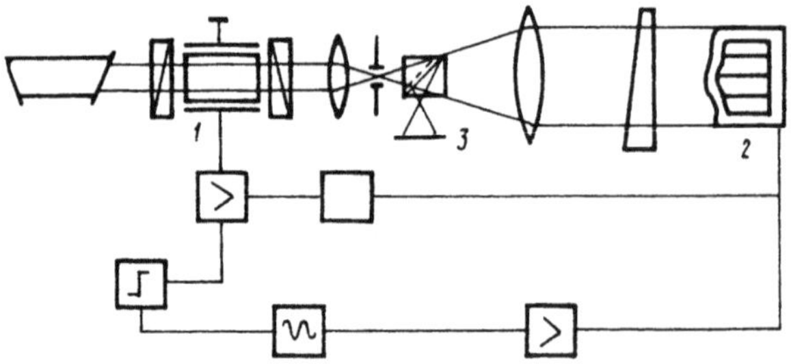

Fig. 4.6 A stroboscopic Fizeau interferometer: *1*—electrooptical modulator, *2*—AM, *3*—registration of interferogram

Interferometric study of the formation of the reflecting surface of the adaptive POE in static and dynamic regimes was performed by using a stroboscopic Fizeau interferometer (Fig. 4.6). The methodology used in interferogram processing allows one to restore the shape of the mirror surface in a given cross section with an error $\lambda/20$ ($\lambda = 0.6328$ μm). The amplitude- and phase-frequency characteristics of the adaptive POE were determined on a vibration stand by using an eddy current sensor of nonstationary displacements whereas the deformation of the actuator was measured with an electronic contact type measurement system.

The characteristic form of the amplitude- and phase-frequency characteristics is shown in Fig. 4.7a, b, whereas the dependence of the elongation of the piezoceramic actuator on the applied voltage in free state and as part of A3—in Fig. 4.7c. The frequency range is 0–5 kHz, and the sensitivity of the actuator in the adaptive POE is 5.3 m/kW, which is five times lower than the sensitivity of the actuator in the free state.

Control of the optical surface by piezoceramic actuators having a hysteresis of the transfer characteristic (Fig. 4.7c) is a challenging task. The uniqueness of formation of a given relief is achieved by 'depolarization' of the piezoceramic actuator by an alternating electric field with a slowly (relative to the oscillation period)

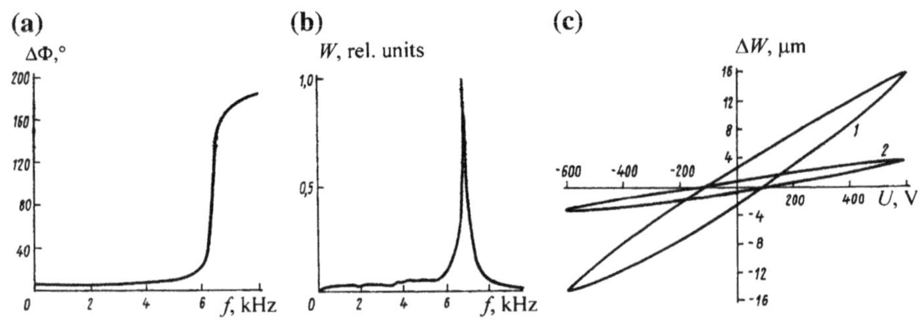

Fig. 4.7 The characteristic form of the amplitude- and phase-frequency characteristics (**a**) and (**b**) is the dependence of the elongation of the piezoceramic actuator on the applied voltage in free state (*1*) and as part of AM (*2*)

decreasing amplitude. This procedure reduces the speed of A3. Local feedback (on the degree of polarization) in the control loop of the piezoceramic actuator allows linearization of its transfer characteristic and maintenance of the performance of A3. For unambiguous and reproducible experimental characteristics of A3 we used the first method of the hysteresis compensation of the piezoceramic actuator.

Consider the results of experimental studies in static regime. Study of the response function of the adaptive POE during the excitation of the actuators by a DC voltage from −600 to +600 V showed that the response function has a bell shape, with the maximum displacement of the mirror surface being localized at the position of the actuator to which the control voltage is applied. Depending on the sign of the control voltage the surface is either convex or concave. Figure 4.8 presents the response function of the mirror surface to the action of the central (1) and lateral (2) actuators at a voltage +600 V. The response functions are similar in the range of −600 to +600 V with an error less than 2 %, which is within the processing error of the interferograms. In the investigated range of the voltages, the maximum displacement of the surface of the adaptive POE linearly depends on the control voltage for the central and lateral actuators (Fig. 4.8b). Thus, the profile of the reflecting surface formed by applying a voltage U to the actuators, can be represented as

$$W(r,\varphi) = \kappa U \Phi(r,\varphi), \tag{4.25}$$

where $\Phi(r,\varphi)$ is the response function normalized to the maximum of the surface displacement and the coefficient k characterizes the dependence of the maximum surface displacement of the adaptive POE on the control voltage. The coefficient κ and the form of the function $\Phi(r,\varphi)$ are not dependent on the voltage applied to the actuator. They are determined by the mutual arrangement of the actuators, their mechanical and piezoelectric properties and the substrate rigidness.

Formation of the profile of the mirror surface by the several actuators was investigated while simultaneously applying identical control voltages from −600 to 600 V to the central and two lateral actuators. These results show that the effect of the actuators on the adaptive POE substrate is additive (Fig. 4.8c), i.e., the reflecting surface resulting from a voltage applied to all actuators is described by the expression

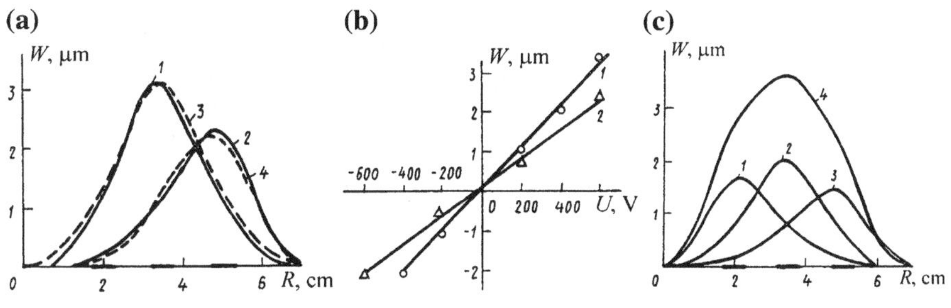

Fig. 4.8 The response function, (*1*, *2*)—experimental and (*3*, *4*)—theoretical, of the mirror surface to the action of the central (*1*) and lateral (*2*) actuators at a voltage +600 V. **a** POE surface max displacement dependence on applied voltage to the central and lateral actuators, **b** surfaces profile of adaptive POE under action of (*1*)—left, (*2*)—central, (*3*)—right actuator, (*4*)—three actuators

$$W(r,\varphi) = \sum_{i=1}^{5} W_i(r,\varphi), \tag{4.26}$$

where $W_i(r,\varphi)$ is the profile of the POE surface formed under the influence of the ith actuator, defined by formula (4.25).

Thus, for a predetermined shape of the mirror surface to be formed in static regime it is necessary to experimentally establish the dependence of the maximum displacement of the reflecting surface on the voltage exciting the actuator and the normalized response function.

For the adaptive POE with a flexible substrate and discrete actuators in static regime the response function is determined by the elastic forces from the unexcited actuators. Upon excitation of the actuators by a time-varying control voltage, the elastic inertial forces arise due to the finite mass of the substrate and actuators, the partial contribution of these forces being determined by the time constant of the control signal.

When forming the surface of a given shape, important is the time interval, where valid are the stationarity and boundary conditions under which the inertial effects become dominant.

Formation of the response function in the dynamics was investigated for a lateral actuator, to which a sinusoidal voltage with an amplitude of 400 V at frequencies 0–3 kHz was applied. Figure 4.9 shows the profiles of the surface, corresponding to the amplitude of the exciting voltage of different frequencies. At frequencies above ~100 Hz, the form of the response function depends on the frequency of the control signal.

In the frequency range up to 2.4 kHz, the reflecting surface at an amplitude of the exciting voltage has a concave shape, the deflection decreases with increasing frequency, and the shape of the surface profile becomes flatter. At a frequency of

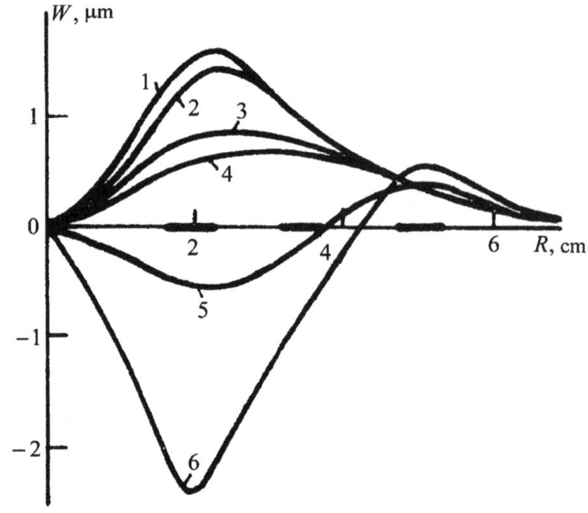

Fig. 4.9 Dynamic response functions of adaptive POE under voltage −400 V applied to the side actuator. Frequency of applied voltage: 0 (*1*), 600 (*2*), 1,800 (*3*), 2,000 (*4*), 2,600 (*5*), 3000 (*6*)

~600 Hz the maximum displacement of the reflecting surface by 13 % is different from the corresponding value in static regime, and at a frequency of 2 kHz this difference amounts to 63 %. By increasing the frequency of the exciting signal from 2.4 to 3 kHz, the shape of the reflecting mirror surface rapidly changes. At these frequencies, oscillations of the adaptive POE substrate are excited, which are similar to the second oscillation mode of a circular plate.

Thus, in static regime the conditions of which are valid in the frequency range of about 100 Hz, the interaction of an actuator with the substrate is reversible in character. In the case of simultaneous action of multiple actuators on the substrate, the surface to be formed is a superposition of the response function of the adaptive POE of each of actuators separately, i.e., to describe the response function, it is necessary to use the linear theory of elasticity and piezoelectric effect.

For the adaptive POE with a continuous reflecting surface and discrete actuators, the fatigue conditions of the substrate thickness in comparison with its radius and smallness of the surface displacements in comparison with the substrate thickness are common. If these conditions are met, the elastic deformation of the substrate by a system of actuators can be considered within the framework of the theory of a thin plate. The deflection of the adaptive POE substrate is described by the equation

$$\Delta\Delta W(r,\varphi) = p(r,\varphi)/D, \ldots \tag{4.27}$$

where $p(r,\varphi)$ is the load per unit area; $D = Eh^3/12(1-\mu^2)$; E is Young's modulus; μ is Poisson's ratio of the material of the substrate; and h is the thickness thereof.

In the case of the rigid fixing of the edge of the adaptive POE, boundary conditions $W|_\Gamma = 0$, $\partial W/\partial n|_\Gamma = 0$ are met. If at a point with the coordinates ρ and θ a transverse point force P acts on the substrate, the solution to (4.3) with the given boundary conditions can be written as

$$\begin{aligned} W(r,\varphi) = {} & (R^2P/16\pi D)\{(1-r^2)(1-\rho^2) + (r^2+\rho^2 \\ & - 2r\rho\cos(\varphi-\theta))\ln[(r^2+\rho^2-2r\rho\cos(\varphi-\theta))/(1 \\ & + r^2\rho^2 - 2r\rho\cos(\varphi-\theta))]\} = (R^2P/16\pi D)\alpha(r,\varphi,\rho,\theta), \end{aligned} \tag{4.28}$$

where R is the radius of the substrate.

Consider the problem of deformation of a circular substrate on the perimeter and resting on N elastic supports under the action of a point force F applied at the location of one of the supports. When a force is applied the surface of the substrate is deformed and reaction forces appear at the locations of the supports. Consequently, the emerging relief can be represented as a superposition of the response function of a thin plate to the action of the force F and the reaction forces from the unexcited actuators. The reaction force R_{ij} arising at the position of the ith support, when a force F_j is applied to the jth support ($i \neq j$), is determined from the system of linear equations

$$(-\alpha_{ik} + 16\pi D\delta_{ik}/R^2K_i)R_{i(\mu)} = \gamma_{ik}R_{i(\mu)} = F_\mu\alpha_{\mu k}, \tag{4.29}$$

where $\alpha_{\mu\gamma} = \alpha(\rho_\mu, \theta_\mu; \rho_\gamma, \theta_\gamma)$ and K_i is the rigidity of the ith actuator.

Thus, surface of the plate formed under the action of a point force F_μ has the form

$$W(r,\varphi) = (R^2 F_\mu / 16\pi D)\left(\alpha_\mu + \sum_{\substack{i=1 \\ i\neq\mu}}^{N} ((\gamma_{ik}^{-1}\alpha_{\mu k})\alpha_i)\right) = F_\mu f_\mu(r,\varphi), \quad (4.30)$$

where $f_\mu(r,\varphi)$ is the response function to the action of a unit force by the μth actuator. When control voltages are applied to all actuators, due to the linearity of the equations of a thin plate the shape of the adaptive POE surface is described by the expression

$$W(r,\varphi) = \sum_{i=1}^{N} F_i f_i(r,\varphi). \quad (4.31)$$

In a piezoceramic actuator the longitudinal deformation is caused by transverse or longitudinal piezoelectric effect. The relationship between electrical and mechanical variables in the case of the transverse piezoelectric effect can be written as:

$$S_1 = s_{11}T_1 + d_{31}E_3, \quad D_3 = d_{31}T_1 + \varepsilon_{33}E_3, \quad (4.32)$$

where S_1 is the longitudinal deformation of the actuator; T_1 is the mechanical stress in the cross section of the actuator; E_3 is the intensity of the transverse electric field in the material of the actuator; D_3 is the electric induction arising in piezoceramics; s_{11} is the elastic flexibility; d_{31} is piezoelectric modulus; and ε_{33} is the permittivity of the actuator material. The actuator force is, therefore, due to the application of a voltage, which is defined by the linear relation

$$F_\mu = K_\mu d_{31} U_\mu l / d(1 + f_\mu(\rho_\mu, \theta_\mu) K_\mu), \quad (4.33)$$

where l and d are its length and width.

Piezoelectric mechanical properties of the actuators can vary. In this case, to determine the response function and the coefficient of proportionality between the actuator force and applied voltage we must introduce for each actuator the parameters of rigidity and piezoelectric module. The presented method for describing the adaptive POE surface takes into account the differences in the properties of the actuators.

A thin plate model describes the properties of formation of the surface profile, which were established experimentally: similarity of the response functions formed by applying different voltages to one actuator, linear dependence of the amplitude of the displacement of the adaptive POE surface on the control voltage applied to the actuator, and additivity of the actuator effect on the substrate.

Consider the problem of the accuracy of the description of the response function using the thin plate model. For a five-element adaptive POE the response functions to the action of the lateral and central actuators were found interferometrically

(Fig. 4.8a, solid line) and by the formulas (Fig. 4.8a, dotted line). For the response function of the central actuator the rms error of approximation is 2.7 % and for the lateral actuator −3.4 % of the maximum deflection of the mirror surface. At a control voltage of +600 V it is 0.06 and 0.1 μm, respectively. Thus, the surface profile of the adaptive POE, which is formed in a given cross section, can be approximated by the response function of a thin plate with an error of about 3 %.

4.3 Accuracy of Correction of Initial Aberration of an Adaptive POE

In correcting the phase distortions of the wavefronts, it is necessary to apply the voltages U_i to the actuators of the adaptive POE [83–86, 109–113, 119, 120]. These voltages cause the development of the 'force' Q_i, providing the formation of the reflecting surface of a given shape. The equation for the surface can be written as $W_a(r,\varphi) = w_0\psi(r,\varphi)$, then the problem of the formation of the reflecting surface of a given shape is reduced to the choice of Q_i, ensuring the best possible approximation $W_a(r,\varphi)$ to the adaptive POE surface being formed. As a measure of approximation we choose the standard deviation

$$\varepsilon^2 = (\pi A^2 w_0^2)^{-1} \int_0^{2\pi}\int_0^{A} (W(r,\varphi) - W_a(r,\varphi))^2 r dr d\varphi,$$

where A is the sub-aperture radius at which the surfaces are conjugated and $W(r,\varphi) = w_0 \sum_{i=1}^{N} Q_i \alpha_i(r,\varphi)$ is the shape of the adaptive POE surface.

The condition of the minimum ε (the least squares method) has the form

$$\beta_{ik} Q_i = \gamma_k, \quad \beta_{ik} = \int_0^{2\pi}\int_0^{A} \alpha_i(r,\varphi)\alpha_k(r,\varphi) r dr d\varphi;$$

$$\gamma_k = w_0 \int_0^{2\pi}\int_0^{A} \psi(r,\varphi)\alpha_k(r,\varphi) r dr d\varphi.$$

Distortion of the intensity distribution of the laser radiation is usually determined by the lower-order wave aberrations. We conducted a simulation of the correction of the first five aberrations of the wavefront by a circle adaptive POE in the form of a plate with a sealed perimeter edge. These included a field curvature $\psi_1(r) = Ar^2$, spherical aberration $\psi_2(r) = Ar^4$, distortion $\psi_3(r,\varphi) = Ar\cos\varphi$, coma $\psi_4(r,\varphi) = Ar^3\cos(\varphi)$, and astigmatism $\psi_5(r,\varphi) = Ar^2\cos^2\varphi$ $(0 \le r \le 1,$ $0 \le \varphi \le 2\pi)$. For the POE in question the possibility of controlling the surface shape is limited because the displacement and slope of the mirror edge are determined by the conditions of its sealing. Reducing the impact of edge sealing on the accuracy of conjugation of the adaptive POE surface with the form of the wavefront aberration is achieved by limiting the diameter of the aperture, on which conjugation was carried out (Fig. 4.9).

Location of actuators in the nodes of a hexagonal grid is typical for adaptive POEs. This configuration is characterized by the number of nodes, N, and the distance between nodes, b. In studying the accuracy of correction of initial aberrations we considered actuator configurations with $N = 7$, 19, 37, and 61; the distance b was varied from a value corresponding to the filling of half the full aperture with actuators to a value corresponding to an almost complete filling of the aperture. The radius of the 'active' aperture of the adaptive POE varied from 1 to 0.6 in the calculations. Figure 4.10 shows the dependences of residual errors of conjugation of the surface of a 37-element adaptive POE with five initial forms of aberrations on the distance between the actuators forming a hexagonal configuration for three values of the active sub-aperture radius. For the aberrations and aperture radii under study advantageous is the location of the actuators at a maximum distance from each other, which corresponds to the most complete filling of the adaptive POE surface by the actuators.

Apparently, when the radius of the aperture is $A \geq 0.6$, the influence of the sealing of the substrate boundary on the shape of the reflecting surface of the adaptive POE is still large enough; to compensate for this effect it is necessary to place some actuators outside the 'active' aperture. This characteristic feature holds true only for the five types of aberrations and hexagonal placement of the actuators. To determine the optimal location of the actuators it is necessary in each case to carry out similar

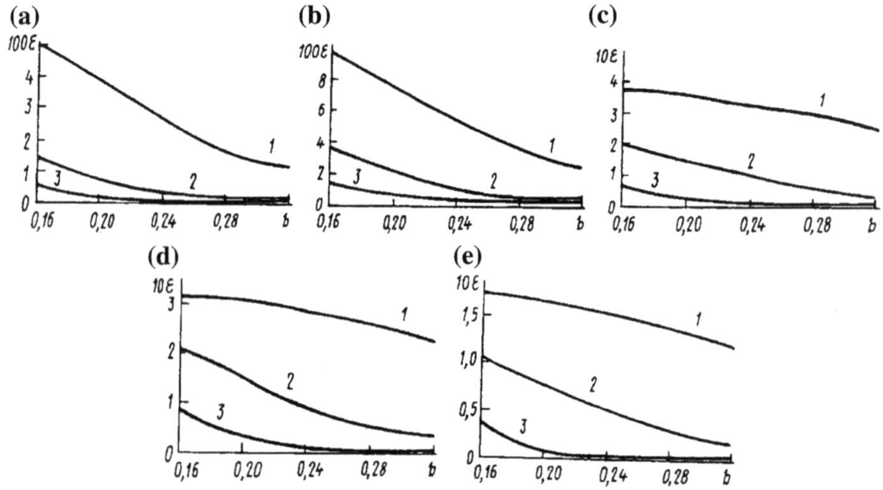

Fig. 4.10 The dependences of residual errors of conjugation of the surface of a 37-element adaptive POE with five initial forms of aberrations on the distance between the actuators forming a hexagonal configuration for three values of the active sub-aperture radius: $A = 1$ (*1*), 0.8 (*2*), 0.6 (*3*)

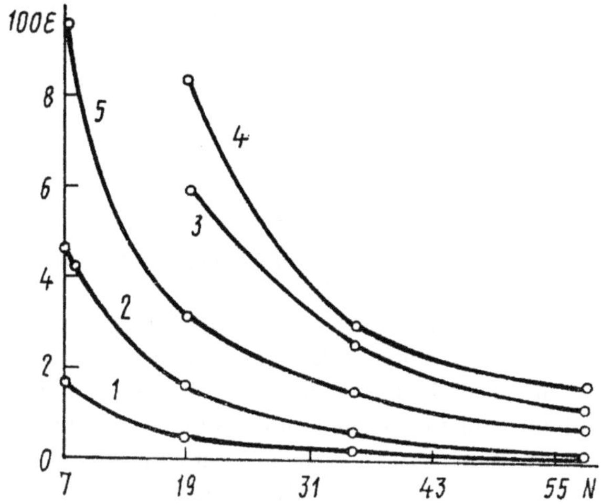

Fig. 4.11 The dependences of the error of conjugation of the surface with five aberrations on the number of the actuators for $A = 0.8$ and number of actuators $N = 7, 19, 37, 61$

calculations with the respective initial data. Figure 4.11 shows the dependences of the error of conjugation of the surface with five aberrations on the number of the actuators, N (aperture $A = 0.8$, location of actuators in each case is optimal). These dependences are conditional, because the number of actuators in the case of hexagonal arrangement can only take fixed values. The general trend of these relationships is a more gradual decrease in A with increasing N, which indicates the inexpediency of an increase in the number of the actuators with a sufficiently smooth distortion compensation when the number N_0 is exceeded ($N_0 \sim 50$–60).

The selected range of the parameters ($N \leq 61,\ A \geq 0.6$), the minimum conjugation errors are $N = 61$ and $A = 0.6$ and amount to $10^{-3} - 10^{-2}$, depending on the type of correctable aberrations. This accuracy corresponds to a high quality of correction of long-wavelength and visible radiations of lasers. At a constant number of the actuators the residual error of conjugation depends on their placement. We considered three configurations of placement of 19 actuators in the adaptive POE (Fig. 4.12). The errors of conjugation of the surface aberrations with the

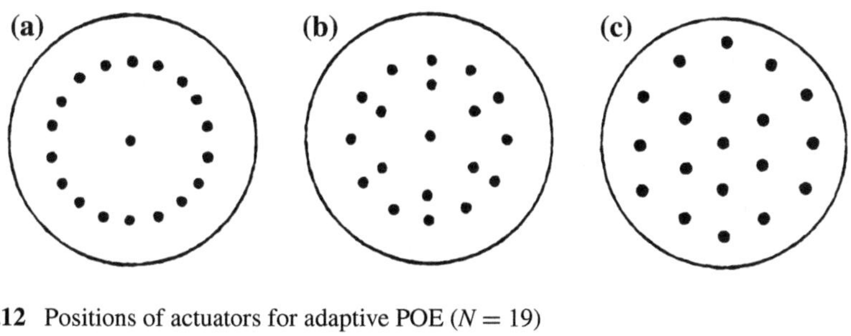

Fig. 4.12 Positions of actuators for adaptive POE ($N = 19$)

Table 4.2 Conjugation errors of the surface aberrations with three optimal in the case of optimal arrangement of the actuators

Configuration	Field curvature	Astigmatism	Coma
рис. 4.12, a	0.001	0.050	0.080
рис. 4.12, b	0.009	0.030	0.060
рис. 4.12, c	0.005	0.035	0.043

three optimal in the case of optimal arrangement of the actuators in each of the configurations are given in the Table 4.2.

Optimal location of the actuators depends on the shape of the corrected aberration. To correct the field curvature, preferable is the location of all the actuators in the adaptive POE on one circle. For the astigmatism to be corrected, the best is the configuration with the actuators located on two circles of different diameters, and in the case of the correction of coma—a hexagonal arrangement of the actuators.

In studying the accuracy of the wavefront distortion corrections the requirements to the adjustment of the optical system, which includes an adaptive POE, are quite important. Consider the conjugation of the surface (hexagonal arrangement of 19 actuators) with the aberration having the field curvature under the misalignment conditions, i.e., when the axis of the radiation incident on the surface of the adaptive POE is shifted relative to the center by a distance δ, varying from 0.01 to 0.3 (the distance between the actuators $b = 0.3$, the aperture $A = 0.6$). Calculations show that even a slight shift of the beam axis relative to the center increases dramatically the error (Fig. 4.13a). The shift by 0.01 leads to a four-fold increase in the error, and the shift by 0.1 increases the error by 30 times.

In the above cases of the correction of the wavefront aberrations we assumed that the optical system was aligned to the normal incidence of radiation on the surface. However, the surface of the adaptive POE can be positioned at an angle to the axis of the incident radiation. In this case, the conjugation accuracy should vary. Figure 4.13b shows the dependence of the error of the correction of astigmatism and field curvature of the adaptive POE with a hexagonal arrangement of 19 actuators on the angle of the radiation incidence on the reflecting surface ($b = 0.48$, $A = 0.6$). The minimal error of conjugation of the surface with the form of the wave aberration is achieved in the case of normal incidence of radiation.

Numerical modeling of the correction of aberrations showed that there is no universal configuration of the arrangement of a given number of actuators, which

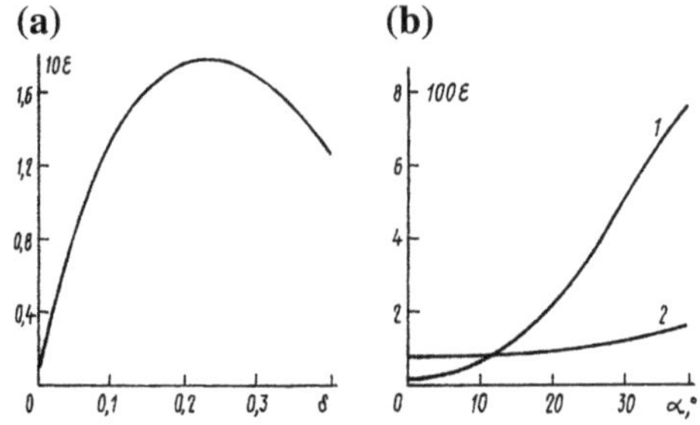

Fig. 4.13 The dependence of the error of the correction of astigmatism and field curvature of the adaptive POE with a hexagonal arrangement of 19 actuators on the angle of the radiation incidence on the reflecting surface ($b = 0.48$, $A = 0.6$)

provides extremely high correction accuracy of a given set of aberrations. For each aberration there exists its own configuration of the optimal number of actuators, ensuring the required correction accuracy for a given sub-aperture. The adaptive POE is centrosymmetric optical element. In an optical system the adaptive POE must be adjusted with high accuracy so as to achieve a desired accuracy of correction. Phase distortion compensation accuracy significantly depends on the angle of incidence of laser radiation on the reflecting surface. Extremely high correction accuracy is achieved at normal incidence of radiation.

Due to the sealing of the perimeter of the adaptive POE substrate, extremely high accuracy of correction of such aberrations as distortion, coma, astigmatism and higher-order aberrations cannot be achieved in principle on the entire aperture.

The study of the response function of the adaptive POE also showed that the formation of the reflecting surface conjugated with wavefront aberrations can be described with high accuracy in terms of a thin plate model. Numerical simulation of the adaptive POE operation allowed us to evaluate the effectiveness of initial aberration correction as a function of the number and location of actuators, the conditions of the laser radiation incidence on the POE surface.

4.4 Adaptive Cooled POEs with Magnetostrictive Actuators of Spring Type

Adaptive optics techniques based on the use of a 'flexible' adaptive POE as a phase corrector [117, 121] are very effective for enhancing the radiation quality of high-power gas lasers. The relief of the reflecting surface, which is conjugate in shape with the wavefront distortions, is formed by deforming the adaptive POE substrate by piezoceramic or magnetostrictive actuators.

Exposure to radiation, wide dynamic range and high spatial frequencies of the aberrations to be corrected impose stringent requirements to adaptive POEs. The use of a forced cooling system—a plate of an adaptive POE—within a flexible substrate (we have chosen a porous structure for thermal stabilization of the mirror surface due to its unique thermal properties) to provide conditions for the optical performance of the reflecting surface leads to an increase in mass and rigidity of the substrate, as well as to an increase in energy spent on its deformation. In this connection, additional requirements arise as to the efficient use of actuators.

For the aberrations to be corrected, use is made of the actuators ensuring the positioning range (10–20)λ with an accuracy $\lambda/20$ (λ is the wavelength of corrected radiation). To solve this problem, promising are spring magnetostrictive actuators (SMAs) based on the Wiedemann effects, which realize relative deformations ~10^{-3} in the frequency band up to 50 nm. The authors of [121] showed the high efficiency of SMAs in the corrector of wavefront slopes. We have studied the cooling of the adaptive POE through its porous structure; the shape of the reflecting surface was formed by nineteen SMAs located on the flexible reflecting substrate of spring type.

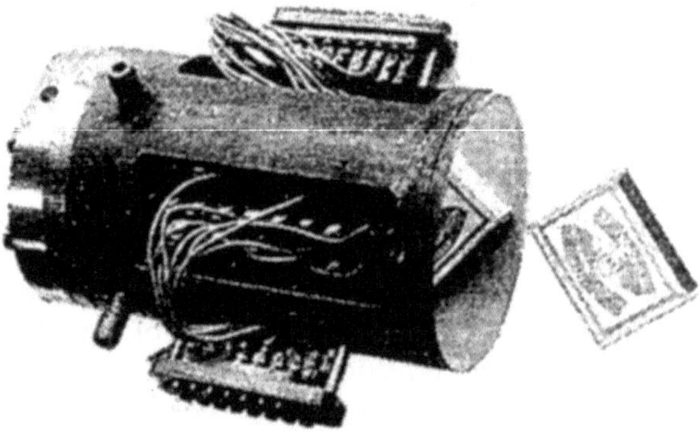

Fig. 4.14 Photo of adaptive POE

The photograph of the POE is shown in Fig. 4.14. Designers were challenged to fabricate a cooled reflecting plate with a substrate having minimal rigidity and thickness, which would ensure effective heat removal from the surface of the adaptive POE and high uniformity of heat and mass transfer in the porous heat exchanger. A relatively large number of control channels, $N = 19$, under conditions of limited dimensions required the solution to the problem of optimal placement of the actuators.

The adaptive POE was a cylinder 98 mm in diameter and 16 cm in height; its scheme is shown in Fig. 4.15. The POE base 2 and the flexible reflecting plate are attached to the housing 1. Between the base and the reflective plate nineteen SMAs 3 are mounted. From the side of the reflective plate the actuators have conical tips 4, the ends of which are glued to the back side of the reflecting plate with an epoxy compound. The actuators are arranged so that the regions of the impact on the flexible plate are located in hexagonal lattice nodes with a period of 16 mm.

Shanks of the actuators are fixed in differential screws 5 of the base, allowing one to adjust statically the shape of the reflecting surface. The full speed of the differential screw ensures the surface deflection of 20 μm. The reflecting plate consists of a copper substrate 6 with cone tips of the actuators, distribution layer 7 and porous copper heat exchanger 8. The distribution layer has a complex channel-hole structure and provides uniformity of mass transfer through the porous heat exchanger. The coolant inlet and outlet are uniform on the periphery of the heat exchanger; the total thickness of the reflecting plate with the heat exchanger is 9 mm. To improve the efficiency of heat removal the reflecting surface is formed directly on the porous structure with the help of curling and polishing of the outer surface with a reflecting layer made of intermetallic Sn_nCu_m.

The POE design makes use of a SMA in the form of a cylindrical bilifar helix made of the 49KF alloy (49 % Fe, 49 % Co, and 2 % V) with an external diameter of 1 cm and height of 14 cm. The SMA operating principle is based on the conversion of torsional deformations by the spring, arising in the cross section of the spring coil upon its magnetization by a combination of circular and axial fields into the longitudinal movement of the spring end. The optimal method for producing an axial field is the transmission of the current through a coil wound on

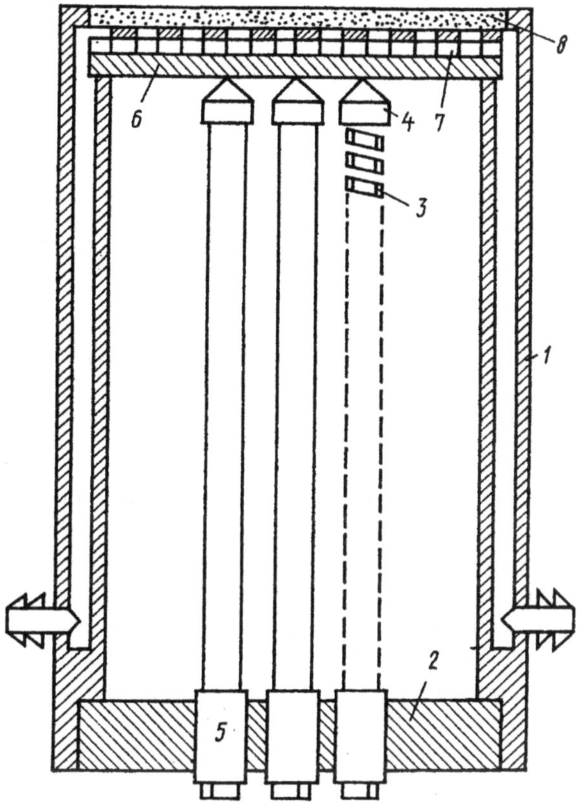

Fig. 4.15 Scheme of adaptive POE

the coils of the spring transformer, with a circular field generated by the current flowing through the coils of the spring transformer. The change of one of the magnetizations leads to a corresponding change in the local swirl angle, which in turn causes the longitudinal movement of the free end of the SMA.

However, the high density of the actuators in the adaptive POE body does not allow one to place axial magnetizationon windings on the SMA. To produce an axial magnetic field, use is made of topological isolation of a bilifar helix. Its unfolding is a closed magnetic circuit consisting of two halves. This circuit of DC magnetic current is possible to implement after placing a permanent magnet in the open circuit. In our case, an axial magnetic field was produced by a permanent magnet made of $SmCo_5$, placed at the end of the spring, which separated the closed magnetic circuit, and the control was carried out by transmitting the control current through the coils of the spring.

Complexity of the reflecting POE plate structure makes it impossible to construct a precise analytical model of its deformation under the action of SMAs. It is therefore of interest to study the adequacy of simpler mathematical models describing the effect of the response function to the action of the actuator (in particular, the models in the approximation of linear theory of elasticity, when the

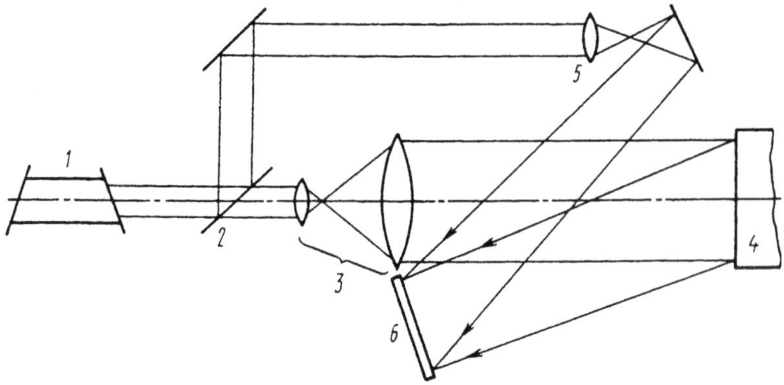

Fig. 4.16 Holographic interferometer optical scheme for POE testing

reflecting plate is considered to be thin and homogeneous, and the applied loads—point-like). For this purpose, we studied experimentally the response function of the adaptive POE, which was then compared with the response function calculated in the approximation of a thin plate.

It was found in the experiments that the maximum deflection of the reflecting surface by a single actuator is no greater than 0.5 μm, which complicates precise interferometric study of the response function. Moreover, the hysteretic nature of the response to the control requires special measures to eliminate the 'zero' interference pattern associated with hysteresis. Therefore, the response function of the POE was measured with a holographic interferometer whose optical scheme is shown in Fig. 4.16. To increase the number of bands on the interferogram and to enhance the relative accuracy of the response function reconstruction, the POE surface was deformed by rotating the differential screw corresponding to the selected control channel.

The interferometer consists of a single-mode He–Ne laser 1, the radiation of which is divided into an object and a reference beam by a beam splitter 2; the object beam is expanded by the telescope 3 and illuminates the object under study 4. Light scattered by the object illuminates the photographic plate 6, onto which a spherical reference wave formed by a microlens 5 is incident. To register the holograms we used VRL photographic plates and the ratio of illuminations created by the object and reference beams was chosen to be 1:3. The holographic measurements were performed by a double exposure method, and the interferogram was recorded in the fringes of infinite width. In measurements we used a POE with a diffusely scattering polished surface of a flexible plate without a reflecting coating.

In reconstructing the response function, the holographic interferogram was applied over the polar coordinate system whose origin coincided with the geometric center of the interferogram. In the chosen coordinate system we built 16 diametrical cross sections of the interferogram with a uniform spacing in azimuth and measured in each section the coordinates of the interference fringes. A closed fringe without any fringes limiting it in space was used as a zero one. In

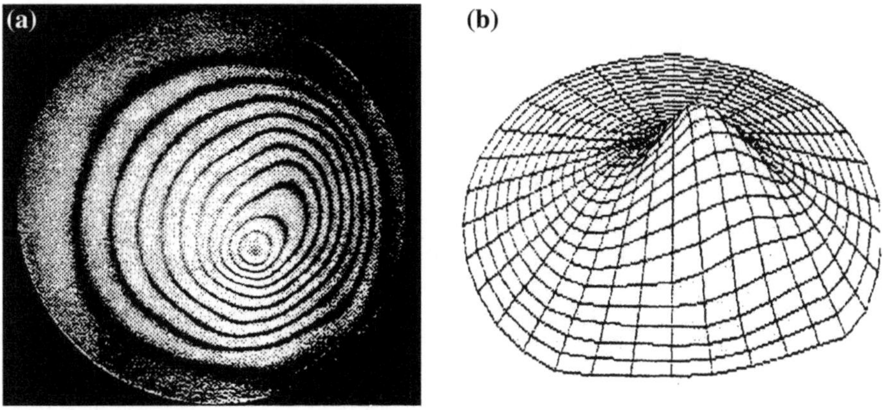

Fig. 4.17 The holographic interferogram of the adaptive POE response to the effects of the peripheral actuator (**a**) and the corresponding three-dimensional image of the response function (**b**)

the measurements we obtained a set of 16 arrays of abscissas and ordinates corresponding to 16 diametrical cross section of the interferogram. The resulting set was then input into the computer, which, with the help of cubic splines followed by smoothing and residue of the slope, constructed a map of normal deviations in a net, which is uniform in radial and azimuthal coordinates. Figure 4.17 shows the holographic interferogram of the response to the effects of the peripheral actuator and the corresponding three-dimensional image of the response function.

To investigate the accuracy of the method we obtained independently several interferograms of the same response function with a different number of interference fringes. They were independently used to reconstruct the relevant maps of normal deviations, the comparison of which showed that, when the number of rings is $N = 6$–11 which is a measure of the measurement error, the normalized standard deviation is $\delta = \int (\varphi - K\psi)^2 dS / \int \varphi^2 dS < 0.05$, $K = \int \varphi\psi dS / \int \varphi^2 dS$; all integrals are taken over the area of the interferogram S; φ and ψ are the independently measured and reconstructed response functions to the action of the same actuator.

In comparing the experimental response functions with the calculated ones in the framework of the mathematical model, we used the representation of the adaptive POE in the form of a plate rigidly fixed to the edge of the deformable by concentrated loads applied to the points of action of the actuators. The model took into account the dimensionless ratio of the rigidity of the deformable plate to the rigidity of the actuator G, defined as the ratio of absolute elongation of the actuator in the free state to the deflection of plates deformed by this actuator. Calculations show a considerable influence of this quantity on the formation of the response function geometry. The experimental study demonstrated that $G \approx 30$–60, i.e., elongation of the actuator in free state is 30–60 times more than the POE deflection with other conditions being equal.

This explains the smallness of the maximum deflection under the action of the actuator. An increase in the maximum deflection is possible if the rigidity of the

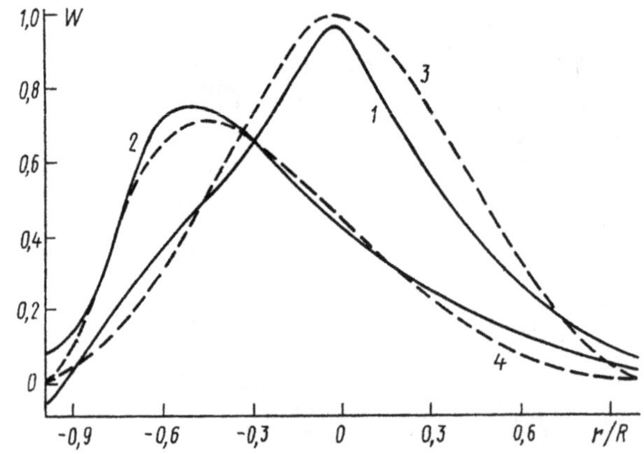

Fig. 4.18 The cross sections of experimental and calculated response function of the adaptive POE under the action of the central and peripheral actuators

flexible reflecting plate is reduced due to an optimal selection of the actuators. The applicability of this layout of A3 is limited by the correction of the static component of the aberration to be corrected; the conjugate surface is formed by using differential screws. Using a dynamic control makes it possible to compensate for small wavefront aberrations caused by thermal deformation of the surface.

Figure 4.18 presents the cross sections of experimental and calculated response function of the adaptive POE under the action of the central and peripheral actuators. The comparison showed that the normalized standard deviation of the calculated response function from the measured one is $\delta \leq 0.05 - 0.1$; the absolute deviation of the theoretical response function from the measured one never exceeds 20 % of the maximum of the measured response function.

Studies have shown high efficiency of holographic interferometry in the studies of the response function and applicability of a simple mathematical model for qualitative assessment of the shape of the response function of a flexible reflecting plate with a complex structure. At the same time, an accurate calculation of the response function of the adaptive POE requires more detailed models.

When designing adaptive POEs it is necessary to ensure the linearity of the response to the control action. This is very important when use is made of adaptive POEs in adaptive optical systems of phase conjugation. However, almost all known drives based piezomaterials have a nonlinear dependence of the elongation of the control action. Hysteresis complicates the control of the surface, and its study is a necessary part of the designing of an adaptive system as a whole; therefore, we studied the hysteresis characteristics of the POEs.

Figure 4.19a presents a hysteresis loop of the current as a function of elongation for the SMA used in the design in free state; the hysteresis is 50 %. In studying the hysteresis of the adaptive POE we observed a qualitative dependence of the hysteresis loop shape of the state in which is the SMA controlling the deflection. The hysteresis loop corresponds to the hysteresis loop of the SMA in free state only when the actuator placed in the adaptive POE was not statically deformed with respect to its free state. Under compression or tension of the SMA, which

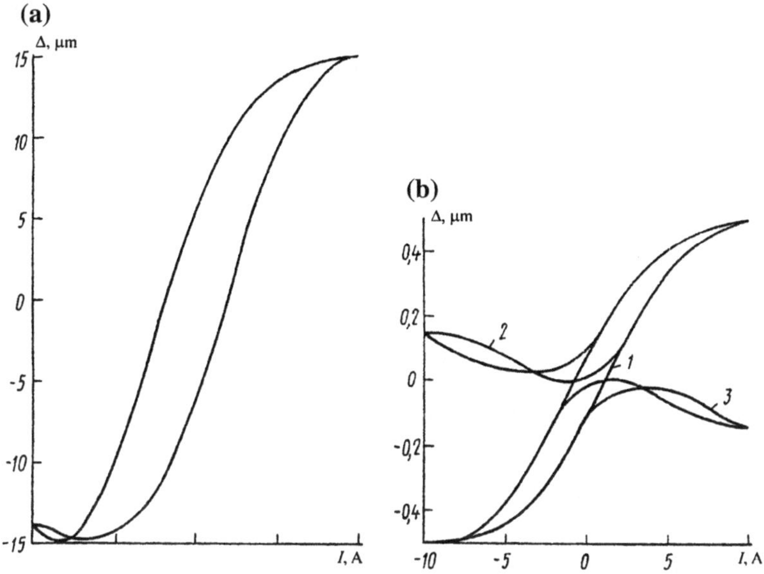

Fig. 4.19 The hysteresis loop of the current as a function of elongation for the SMA used in the design in free state (**a**); the hysteresis loop corresponds to the hysteresis loop of the SMA in free state only when the actuator placed in the adaptive POE was not statically deformed (*1*), under expansion (*2*) and under compression or tension (*3*) (**b**)

was carried out by the differential adjusting screw, the hysteresis loop is distorted and acquires a characteristic appearance of the 'butterfly'. A qualitative change in the hysteresis characteristics under static compression or tension of the SMA suggests the need for accurate matching of the rigidity of the flexible plate with the rigidity of the SMA, as well as the need for special measures to linearize the SMA characteristics.

References

1. V.V. Apollonov, S.A. Chetkin, A.M. Prokhorov, S.N. Temnov, *Lectures on Ernst Abel Conference* (Jena, GDR, 1987)
2. V.V. Apollonov, A.S. Brynskih, S.A. Chetkin, A.M. Prokhorov, S.N. Temnov, *Lectures on Ernst Abel Conference* (GDR, Jena, 1987)
3. V.V. Apollonov, S.A. Chetkin, *Lectures on Ernst Abel Conference* (GDR, Jena, 1987)
4. V.V. Apollonov, S.A. Chetkin, A.M. Prokhorov, S.N. Temnov, *18-ECJIM* (Czechoslovakia, Prague, 1987)
5. V.V. Apollonov, A.S. Brynskih, S.A. Chetkin, A.M. Prokhorov, S.N. Temnov, *18-ECJIM* (Czechoslovakia, Prague, 1987)
6. V.V. Apollonov, G.I. Babayants, I.V. Goncharenko, A.M. Prokhorov, C.A. Chetkin, E.V. Khristyan, in *Proceedings of All Union Conference "Non-Traditional Materials"*, USSR, Moscow, 1987

7. V.V. Apollonov, S.A. Chetkin, A.M. Prokhorov, S.N. Temnov, in *Proceedings of All Union Conference "Non-traditional materials"*, USSR, Moscow, 1987
8. V.V. Apollonov, S.A. Chetkin, A.M. Prokhorov, S.N. Temnov, Preprint GPI RAS **231** (1987)
9. V.V. Apollonov, A.I. Ivlutin, V.N. Motorin, A.M. Prokhorov, V.N. Kharchenko, C.A. Chetkin, Patent No. 1592711 (1987)
10. V.V. Apollonov, I.V. Goncharenko, A.M. Prokhorov , C.A. Chetkin, E.V. Khristyan, A.A. Shubin, Patent No. 1459570 (1987)
11. V.V. Apollonov, I.V. Goncharenko, A.M. Prokhorov , C.A. Chetkin, E.V. Khristyan, A.A. Shubin, Patent No. 4250382 (1987)
12. V.V. Apollonov, N.N. Zubkov, V.N. Motorin, A.M. Prokhorov, V.N. Kharchenko, C.A. Chetkin, in *Proceedings of Forum, Heat-Mass Exchange*, USSR, Minsk, 1988
13. V.V. Apollonov, V.N. Motorin, V.N. Kharchenko, in *Proceedings of Forum, Heat-Mass Exchange*, USSR, Minsk, 1988
14. V.V. Apollonov, I.V. Goncharenko, V.N. Motorin, A.M. Prokhorov, V.N. Kharchenko, C.A. Chetkin, in *Proceedings of Forum, Heat-Mass Exchange*, USSR, Minsk, 1988
15. V.V. Apollonov, S.A. Chetkin,V.N. Kravchenko, A.M. Prokhorov, V.N. Motorin , N.N. Zubkov, in *1st World Conference on Experimental Heat Transfer, Fluid Mechanics and Thermodynamics*, Dubrovnik, Yugoslavia, 1988
16. V.V. Apollonov, S.A. Chetkin, A.M. Prokhorov, S.N. Temnov, V.A. Khmara, Letters to JTP **14**, 236 (1988)
17. V.V. Apollonov, S.A. Chetkin, I.V. Goncharenko , V.A. Khmava, A.M. Prokhorov, in *1st World Conference on Experimental Heat Transfer, Fluid Mechanics and Thermodynamics*, Dubrovnik, Yugoslavia, 1988
18. V.V. Apollonov, C.A. Chetkin, A.M. Prokhorov, *Lasers 1988* (Nevada, USA, 1988)
19. V.V. Apollonov, Lasers 1988. Lake Tahoe, Nevada, USA (1988)
20. V.V. Apollonov, C.A. Chetkin, Preprint GPI RAS No. 65 (1988)
21. V.V. Apollonov, S.I. Zienko, C.A. Chetkin, DTE, 12, (1988)
22. V.A. Alekseev, V.V. Apollonov, C.A. Chetkin, Y.V. Danchenko, V.N. Egorov, S.V. Kulakov, Patent No. 4250395 (1988)
23. V.V. Apollonov, N.N. Zubkov, V.N. Motorin, A.M. Prokhorov, V.N. Kharchenko, C.A. Chetkin, Patent No. 4376825 (1988)
24. V.V. Apollonov, I.V. Goncharenko, N.N. Zubkov, V.N. Motorin, A.M. Prokhorov, V.N. Kharchenko, C.A. Chetkin, Patent No. 4376814 (1988)
25. V.N. Antciferov, V.A. Alekseev, V.V. Apollonov, C.A. Chetkin, Y.V. Danchenko, V.N. Egorov, S.V. Kulakov, Patent No. 289258 (1988)
26. V.V. Apollonov, N.N. Zubkov, V.N. Motorin, A.M. Prokhorov, V.N. Kharchenko, C.A. Chetkin, Patent No. 4368523 (1987)
27. V.V. Apollonov, S.A. Chetkin, S.N. Temnov, QE **15**, 2578 (1988)
28. V.V. Apollonov, N.N. Zubkov, V.N. Motorin, A.M. Prokhorov, V.N. Kharchenko, C.A. Chetkin, *Experimental Heat Transfer Fluid Mechanics and Termodynamics* (Elsevier, 1988), p. 882
29. V.V. Apollonov, C.A. Chetkin, V.N. Kharchenko, V.N. Motorin, A.M. Prokhorov, N.N. Zubkov, *Experimental Heat Transfer Fluid Mechanics and Termodynamics* (Elsevier, 1988), p. 915
30. V.V. Apollonov, A.M. Prokhorov, S.A. Chetkin, S.N. Temnov, Lett. JTP **14**, 3 (1988)
31. V.V. Apollonov, S.A. Chetkin, V.F. Goncharov, N.P. Zubkov, A.I. Ivlutin, V.N. Motorin, Patent No. 1510560 (1988)
32. S.I. Derzhavin, Cand. Diss., GPI RAS, Moscow (1988)
33. V.V. Apollonov, S.A. Chetkin, V.F. Goncharov, N.P. Zubkov, A.I. Ivlutin, V.N. Motorin, Patent No. 1544037 (1988)
34. V.V. Apollonov, S.I. Zienko, S.V. Muraviev, Devices and technique for experiments. No. 3 (1989)
35. V.V. Apollonov, I.G. Odnorozhenko, A.F. Sokolova, Patent No. 304164 (1989)
36. V.V. Apollonov, E.A. Ivanova, A.M. Prokhorov, S.A. Chetkin, Lett. JTP **15**, 22 (1989)
37. V.V. Apollonov, I.G. Odnorozhenko, A.F. Sokolova, Patent Patent No. 322187 (1989)
38. V.V. Apollonov, S.I. Zienko, S.V. Muraviev, Devices and technique for experiments. No. 1 (1990)

39. V.N. Antciferov, V.V. Apollonov, C.A. Chetkin, Preprint GPI RAS No. 66 (1988)
40. V.V. Apollonov, E.A. Ivanova, A.M. Prokhorov, S.A. Chetkin, Patent No. 1753443 (1989)
41. V.V. Apollonov, S.A. Chetkin, A.M. Prokhorov, V.N. Kharchenko, Lett. JTP **15**, 3 (1989)
42. V.V. Apollonov, E.A. Ivanova, A.M. Prokhorov, S.A. Chetkin, S.N. Temnov, J. Exp. Tech. Phys. GDR **37**, 41 (1989)
43. V.V. Apollonov, V.I. Borodin, S.A. Chetkin, E.A. Ivanova, A.M. Prokhorov, S.I. Zienko, S.V. Muraviev, QE **16**, 386 (1989)
44. V.V. Apollonov, L.M. Ostrovskaya, V.N. Rodin, A.A. Shubin, V.A. Shurygin, Patent No. 288490 (1989)
45. V.V. Apollonov, S.A. Chetkin, A.M. Prokhorov, *Laser Damage Symposium* (Colorado, USA, 1989)
46. V.V. Apollonov, *Lasers 1989* (New Orleans, Louisiana, USA, 1989)
47. V.V. Apollonov, S.A. Chetkin, V.F. Goncharov, N.P. Zubkov, A.I. Ivlutin, V.N. Motorin, Patent No. 1672896 (1989)
48. V.V. Apollonov, S.A. Chetkin, V.F. Goncharov, N.P. Zubkov, A.I. Ivlutin, V.N. Motorin, Patent No. 1592711 (1990)
49. V.V. Apollonov, F.M. Prokhorov, V.A. Shmakov, G.I. Babayants, M.V. Gartman, Lett. JTP **16**, 2 (1990)
50. V.V. Apollonov, F.M. Prokhorov, V.S. Kolesov, V.A. Shmakov, V.V. Alksnin, Lett. JTP **16**, 2 (1990)
51. V.V. Apollonov, S.A. Chetkin, E.A. Ivanova, F.M. Prokhorov, QE **17**, 786 (1990)
52. V.V. Apollonov, V.N. Kharchenko, A.M. Prokhorov, N.P. Poluektov, *IHTC-9* (Jerusalem, 1990), p. 51
53. V.V. Apollonov, V.I. Borodin, S.A. Chetkin, E.A. Ivanova, A.M. Prokhorov, QE **17**, 1496 (1990)
54. V.V. Apollonov, S.A. Chetkin, E.A. Ivanova, F.M. Prokhorov, QE **17**, 1509 (1990)
55. V.V. Apollonov, S.A. Chetkin, E.A. Ivanova, F.M. Prokhorov, QE **17**, 1642 (1990)
56. V.V. Apollonov, I.G. Odnorozhenko, A.F. Sokolova, Patent No. 330267 (1990)
57. V.V. Apollonov, D.V. Artemov, V.I. Kislov, F.M. Prokhorov, QE **18**, 359 (1991)
58. V.V. Apollonov, S.A. Chetkin, E.A. Ivanova, A.M. Prokhorov, in *Proceedings of the International Conference of Lasers 1990*, Nevada, USA, 1990
59. V.V. Apollonov, S.A. Chetkin, E.A. Ivanova, A.M. Prokhorov, G.V. Vdovin, in *Proceedings of he International. Congress on Optical Science and Engineering*, Hague, Netherlands, 1990
60. V.V. Apollonov, S.A. Chetkin, V.I. Kislov, G.V. Vdovin, in *Proceedings of he International. Congress on Optical Science and Engineering*, Hague, Netherlands, 1990
61. V.V. Apollonov, G.V. Vdovin, V.I. Kislov, S.A. Chetkin, *V Seminar on Solar Instruments* (Irkutsk, 1990)
62. V.I. Aksinin, V.V. Apollonov, V.I. Borodin, A.S. Brynskikh, S.A. Chetkin, S.V. Murave'v, V.V. Ostanin, G.V. Vdovin, Sens. Actuators A **21**, (1990)
63. V.V. Apollonov, G.V. Vdovin, S.A. Chetkin, Sens. Actuators A **21** (1990)
64. V.V. Apollonov, G.V. Vdovin, V.I. Kislov, S.A. Chetkin, Patent No. 20173136 (1990)
65. V.V. Apollonov, *Lasers 1991* (San Diego, California, USA, 1991)
66. Sholev S.F., Cand. Diss., Moscow, Almaz (1991)
67. V.V. Apollonov, S.A. Chetkin, G.V. Vdovin, *Lasers 1991* (California, USA, 1991)
68. V.V. Apollonov, G.V. Vdovin, V.I. Kislov, S.A. Chetkin, QE **18**, 358 (1991)
69. V.V. Apollonov, A.M. Prokhorov, V.A. Shmakov, Lett. JTP **17** (1991)
70. V.V. Apollonov, G.V. Vdovin, V.I. Kislov, S.A. Chetkin, QE **18**, 1367 (1991)
71. V.V. Apollonov, G.V. Vdovin, V.I. Kislov, S.A. Chetkin, QE **19**, 596 (1992)
72. V.V. Apollonov, S.A. Chetkin, A.M. Prokhorov, in *International Seminar on Advanced Optical Technology*, GDR, Drezden, 1992
73. V.V. Apollonov, *Lasers 1992* (Houston, Texas, USA, 1992)
74. V.A. Shurygin V.A., Cand. Diss., Moscow, Almaz (1992)

75. V.V. Apollonov, V.I. Kislov, D.V. Artemov, QE **20**, 1203 (1993)
76. V.V. Apollonov, A.M. Prokhorov, V.A. Shmakov, *Lasers 1993*, (Nevada, USA, 1993)
77. S.A. Savranskii, Cand. Diss., SIIRO, Moscow (1993)
78. V.V. Apollonov, V.I. Kislov, D.V. Artemov, *Lasers 1994* (Quebec, Canada, 1994)
79. V.V. Apollonov, D.V. Artemov, V.I. Kislov, A.M. Prokhorov, QE **21**, 577 (1994)
80. V.V. Apollonov, in *6th International Symposium on Advanced Nuclear Technology*, Mito, Ibaraki, Japan, 1994
81. V.V. Apollonov, G.V. Vdovin, V.I. Kislov, S.A. Chetkin, Patent No. 2017313 (1994)
82. V.I. Aksinin, V.V. Apollonov, G.V. Vdovin, V.I. Kislov, S.A. Chetkin, Patent No. 2047880 (1995)
83. V.V. Apollonov, A.B. Egorov, Y.L. Kalachev, *Lasers 1995* (Charleston, South Carolina, USA, 1995)
84. V.V. Apollonov, G.I. Babayants, S.I. Derzhavin, V.V. Kuzminov, D.A. Mashkovsky, A.M. Prokhorov, Y.P. Koval, Patent No. 2130221(1996)
85. V.V. Apollonov, V.V. Kijko, V.I. Kislov, *Lasers 1996* (Portland, Oregon, USA, 1996)
86. V.V. Apollonov, A.B. Egorov, Y.L. Kalatchev, V.V. Kijko, V.I. Kislov, *Lasers 1996* (Oregon, USA, 1996)
87. V.V. Apollonov, G.I. Babayants, Y.P. Voinov, A.M. Prokhorov, E.V. Khristyan, B.Sh. Kishmakhov, Patent No. 1277854 (1985)
88. V.V. Apollonov, G.I. Babayants, P.I. Bystrov, V.F. Goncharov, A.M. Prokhorov, Patent No. 244286 (1985)
89. V.V. Apollonov, A.I. Ivlutin, V.N. Motorin, A.M. Prokhorov, V.N. Kharchenko, C.A. Chetkin, Patent No. 1278565 (1985)
90. V.V. Apollonov, I.V. Goncharenko, A.M. Prokhorov, C.A. Chetkin, E.V. Khristyan, A.A. Shubin, Patent No. 1398717 (1985)
91. V.V. Apollonov, I.V. Goncharenko, A.M. Prokhorov, C.A. Chetkin, E.V. Khristyan, A.A. Shubin, Patent No. 1301260 (1986)
92. V.V. Apollonov, A.S. Brynskih, S.A. Chetkin, A.M. Prokhorov, S.N. Temnov, Patent No. 240743 (1986)
93. V.V. Apollonov, V.I. Borodin, A.S. Brynskih, S.A. Chetkin, A.M. Prokhorov, S.N. Temnov, Patent No. 234217 (1986)
94. V.V. Apollonov, V.I. Borodin, A.S. Brynskih, S.A. Chetkin, V.V. Ostanin, A.M. Prokhorov, A.M. Temnov, Patent No. 234216 (1986)
95. V.V. Apollonov, V.A. Alekseev, S.V. Bilibin, Y.V. Danchenko, S.A. Chetkin, A.P. Kunevich, A.M. Prokhorov, Patent No. 241100 (1986)
96. V.N. Antciferov, V.V. Apollonov, V.A. Alekseev, S.V. Bilibin, Y.V. Danchenko, S.A. Chetkin, A.P. Kunevich, A.M. Prokhorov, Patent No. 261206 (1986)
97. V.V. Apollonov, A.S. Brynskii, M.S. Granovskii, L.M. Ostrovskaya, A.M. Prokhorov, C.A. Chetkin, V.N. Rodin, M.I. Tsipin, A.A. Shubin, S.F. Sholev, Patent No. 251080 (1987)
98. V.V. Apollonov, M.S. Granovskii, N.N. Zubkov, V.N. Motorin, L.M. Ostrovskaya, V.N. Rodin, A.M. Prokhorov, V.N. Kharchenko, C.A. Chetkin, Patent No. 1455217 (1987)
99. V.V. Apollonov, B.F. Belov, V.N. Karmatskii, L.M. Ostrovskaya, A.M. Prokhorov, C.A. Chetkin, V.N. Rodin, M.I. Tsipin, S.F. Shevakin, S.F. Sholev, Patent No. 202719 (1987)
100. V.V. Apollonov, M.S. Granovskii, N.N. Zubkov, V.N. Motorin, A.M. Prokhorov, V.N. Kharchenko, C.A. Chetkin, Patent No. 4198992 (1987)
101. V.V. Apollonov, V.I. Andryushin, N.N. Zubkov, V.N. Motorin, A.M. Prokhorov, V.N. Kharchenko, C.A. Chetkin, A.A. Shubin, V.A. Shurygin, Patent No. 4249488 (1987)
102. V.V. Apollonov, A.I. Ivlutin, V.N. Motorin, A.M. Prokhorov, V.N. Kharchenko, C.A. Chetkin, Patent No. 4212816 (1987)
103. V.V. Apollonov, A.I. Ivlutin, V.N. Motorin, A.M. Prokhorov, V.N. Kharchenko, C.A. Chetkin, Patent No. 4227164 (1987)
104. V.V. Apollonov, V.I. Andryushin, N.N. Zubkov, V.N. Motorin, A.M. Prokhorov, C.A. Chetkin, A.A. Shubin, V.A. Shurygin, Patent No. 288490 (1987)
105. V.V. Apollonov, Y.V. Danchenko, S.A. Chetkin, A.M. Prokhorov, Patent No. 261206 (1987)

106. L.S. Tsesnek, O.V. Sorokin, A.A. Zolotukhin, *Metal Mirrors* (Mashinostroenie, Moscow, 1983)
107. A.G. Safronov, Cand. Diss., GPI RAS, Moscow (1996)
108. V.V. Apollonov, A.I. Barchukov, L.T. Kir'yanova, L.M. Ostrovskaya, A.M. Prokhorov, V.N. Rodin, O.S. Serebryannikova, M.I. Tsypin, FIAN Report, Moscow (1977)
109. V.V. Apollonov, G.V. Vdovin, V.I. Kislov, S.A. Chetkin, Patent No. 1753443 (1996)
110. V.V. Apollonov, S.A. Chetkin, V.F. Goncharov, E.V. Khristyan, A.A. Shubin, Patent No. 1459570 (1996)
111. Apollonov V.V, I.G. Odnorozhenko, A.F. Sokolova , Patent No. 1459570 (1996)
112. V.V. Apollonov, S.A. Chetkin, V.F. Goncharov, N.P. Zubkov, A.I. Ivlutin, V.N. Motorin, Patent No. 1544037 (1996)
113. V.V. Apollonov, S.A. Chetkin, V.F. Goncharov, N.P. Zubkov, A.I. Ivlutin, V.N. Motorin, Patent No. 1510560 (1996)
114. V.V. Apollonov, I.G. Odnorozhenko, A.F. Sokolova, Patent No. 322187 (1996)
115. V.V. Apollonov, D.V. Artemov, V.I. Kislov, A.M. Prokhorov, QE **23**, 1081 (1996)
116. V.V. Apollonov, V.V. Kijko, V.I. Kislov, F.B. Egorov, V.I. Aksinin. Patent No. 119588 (1996)
117. V.V. Apollonov, E.P. Bochkar, V.Y. Zaslavskii, V.Y. Khomich, Preprint FIAN (1978)
118. V.V. Apollonov, *Innovations Through Reforms* (MCNTI, Moscow, 1998)
119. V.V. Apollonov, S.I. Derzhavin, V.V. Kuzminov, D.A. Mashkovsky, A.M. Prokhorov, Y.P. Koval, *Lasers 1996* (Oregon, USA, 1996)
120. V.V. Apollonov, V.V. Kijko, V.I. Kislov, QE **23**, 999 (1996)
121. V.V. Apollonov, V.P. Malyavin, V.N. Feofilaktov, *Lasers 1997* (New Orleans, Louisiana, 1997)

Chapter 5
Analytical Model of an Adaptive POE in the Form of a Thin Plate with Discrete Actuators

This part will provide a model of an adaptive POE in the form of a thin plate with discrete piezoactuators. The model is based on the provisions of the thin plate theory. We found analytical expressions for the response function of the adaptive POE, the range of displacements of the substrate under the action of the actuator, the energy conversion efficiency of the control signal to the energy of elastic deformation of the substrate. The results obtained in the framework of the response function model are in good agreement with the experimental response function of a 19-element cooled POE with magnetostrictive actuators. Presented are the results of the numerical simulation of the actuator operation in the adaptive POE. Generation of aberration-free fluxes of coherent radiation is a challenge in power optics. One of the promising methods for solving the problem is the use of adaptive POEs consisting of a phase distortion corrector, a wavefront sensor and a control system. If the phase distortions, which must be compensated for, are smooth, the high quality correction can be obtained by using adaptive POEs with a continuously deformable surface. However, due to non-locality and non-orthogonality of the response function such a correction is associated with significant difficulties. Furthermore, there is no analytical model with a continuously deformable surface that would describe the work of the adaptive POE and predict its characteristics, such as the form of the response function and the range of displacements of the substrate under the action of actuators in the design of this or that POE.

In light of the above-said, the aim of the work, which consists in the analytical description of the adaptive POE in the form of a thin plate and containing discrete piezoelectric or magnetostrictive actuators, is urgent. The thin plate theory is an effective tool for describing the elastic interaction of the substrate and a system of discrete actuators. The displacement of the reflecting surface $w(\mathbf{r})$ by applying a distributed power load $p(\mathbf{r})$ to the a thin plate-substrate satisfies the relationship.

$$\Delta w(\mathbf{r}) = p(\mathbf{r})/D \tag{5.1}$$

with the boundary conditions corresponding to the sealing of the mirror edge. In (5.1) D is the flexural rigidity of the substrate. Consider a circular POE with N actuators and a sealed edge. The control signal is applied to the actuator with the

V.V. Apollonov, *High-Power Optics*, Springer Series in Optical Sciences 192,
DOI 10.1007/978-3-319-10753-0_5

number k and acts with a force P on the substrate, and at the locations of other actuators the reaction forces Pq_i arise, where i is the number of the actuator. In this case, the displacement of the substrate, which is a solution of (5.1), has the form

$$w_k(r,\varphi) = (PR^2/16\pi D)\left[\alpha_k(r,\varphi) + \sum_{i=1, i\neq k}^{N} q_i\alpha_i(r,\varphi)\right], \tag{5.2}$$

where R is the radius of the substrate; $\alpha_j(r,\varphi) = (1-r^2)(1-\rho_j^2) + [r^2+\rho_j^2 - 2r\rho_j \cos(\varphi-\theta_j)]\ln\{[r^2+\rho_j^2 - 2r\rho_j\cos(\varphi-\theta_j)]/[1+r^2\rho_j^2 - 2r\rho_j\cos(\varphi-\theta_j)]\}$; and ρ_j and θ_j are the coordinates of the jth actuator.

Since the elongation of the actuator coincides with the displacement of the substrate at the point of its location, and the force exerted on the actuator by the substrate is equal in magnitude and opposite in direction to the force exerted on the substrate by the actuator, the relative reaction forces q_i are found from the system of linear equations

$$\sum_{i=1, i\neq k}^{N} a_{ij}q_i = -\alpha_j(\rho_k,\theta_k) \quad (j=1,\ldots,N;\ j\neq k), \tag{5.3}$$

where $a_{ij} = \alpha_i(\rho_j,\theta_j) + \delta_{ij}G$; δ_{ij} is the Kronecker delta-function; and $G = 16\pi D/R^2K_a$ (K_a is the rigidity of the actuator) characterizes the ratio of the substrate rigidity to the actuator rigidity.

To control the shape of the reflecting surface of the adaptive POE, use is made of piezoelectric and magnetostrictive actuators. Equations relating the electrical and mechanical variables in the case of a piezoelectric material, as well as magnetic and mechanical variables in the case of a magnetostrictive material are given in the appendixes.

We denote the elongation of the actuator controlled by a given signal in free state by Δl_0 (the relationship Δl_0 with the signal is presented in the appendix). Then, the actuator elongation in the adaptive POE has the form

$$\Delta l = P/K_a + \Delta l_0. \tag{5.4}$$

On the other hand, the elongation with the displacement w_k of the actuator substrate (5.2) is:

$$\begin{aligned} &-P/K_a + \Delta l_0 = [PR^2/ \\ &+ \sum_{i=1, i\neq k}^{N} q_i\alpha_i(\rho_k,\theta_k)]\,R^2/16\pi D \end{aligned} \tag{5.5}$$

and, therefore, the force in A3,

$$P = \Delta l_0/\{K_a^{-1} + [\alpha_k(\rho_k,\theta_k) + \sum_{i=1, i\neq k}^{N} q_i\alpha_i(\rho_k,\theta_k)]R^2/16\pi D\}. \tag{5.6}$$

By substituting (5.6) into (5.2), we obtain an analytical expression for the response function to the action of the kth actuator:

$$w_k(r,\varphi) = \Delta l_0 f_k(r,\varphi)/f_k(\rho_k,\theta_k) + G, \quad (5.7)$$

where $f_k(r,\varphi) = \alpha_k(r,\varphi) + \sum_{i=1,i\neq k} q_i\alpha_i(r,\varphi)$ is the normalized response function, which determines the shape of the reflecting surface of A3 and depends on the number and mutual arrangement of actuators, as well as on the relative rigidity G. The amplitude of the substrate displacement under the action of the actuator $\Delta l_0/[f_k(\rho_k,\theta_k) + G]$ depends on the number and location of the actuators, on the characteristics of the materials used to fabricate the actuators and the substrate and on the control signal. As an example, Fig. 5.1 shows the experimental response functions of a 19-element cooled mirror with magnetostrictive actuators and the response functions of a thin plate with the parameter G, corresponding to the minimum standard deviation of the experimental curves from the calculated ones.

The rms error in the approximation of experimental response functions by the theoretical ones is 0.06–0.10 of the standard deviation of the experimental response function from the plane. The approximation error is associated with the design features of the adaptive POE whose substrate is made in the form of a thick enough plate having a complex structure because of the cooling system. Nevertheless, the model description gives good results even away from the strict

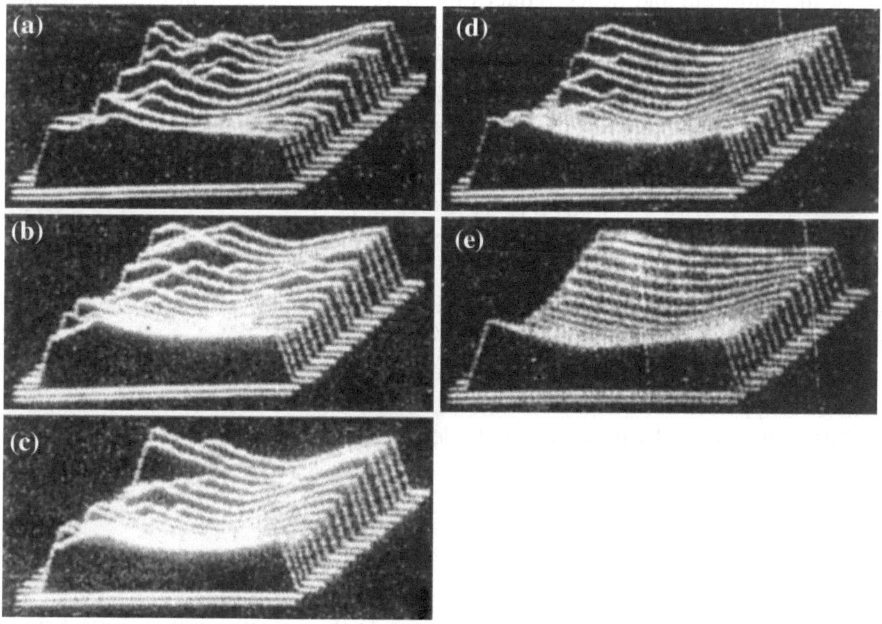

Fig. 5.1 The experimental response functions of a 19-element cooled mirror with magnetostrictive actuators (**a**) and the response functions of a thin plate, corresponding to four steps of iteration procedure (**b**–**e**)

conditions of its applicability; in particular, the limitation regarding the thickness of the substrate is not rigid.

The efficiency of the actuator operation is characterized by the ratio of the actuator elongation in the POE to the actuator elongation in free state and the energy conversion efficiency (ECE) of the control signal to the energy of deformation of the substrate. Elongation of the actuator in the adaptive POE with respect to elongation of the actuator in free state, as follows from (5.4)–(5.6), is

$$\Delta l/\Delta l_0 = 1 - G/[G + f_k(\rho_k, \theta_k)]. \tag{5.8}$$

Thus, the range of displacements of the actuator when it is installed in the POE is determined by the dimensionless parameter G, the position of the actuator with respect to the center of the surface, the configuration of the placement of actuators and their total number. The dependences $\Delta l/\Delta l_0(G)$ shows for the POE with one actuator at different positions of the actuator with respect to the center. An increase in G reduces the range of the actuator displacements, removal of the actuator from the center also reduces $\Delta l/\Delta l_0$. The dependences $\Delta l/\Delta l_0(G)$ for the actuators of the 19-element adaptive POE located at the nodes of the hexagonal grid with a spacing $b = 0.3R$. For actuators arranged at an equal distance from the center the dependences are equal; with increasing distance from the center, as in the case of a single actuator, $\Delta l/\Delta l_0$ is reduced. Comparison of the dependences shows that increasing the number of the actuators increases the effective rigidity of the substrate and hence reduces the range of displacement of an individual actuator.

The energy conversion efficiency of the control signal of the actuator can be characterized by the ECE equal to the ration of the strain energy of the substrate relative to the total energy of the actuators and the substrate:

$$\eta = U_0 \Big/ \left(\sum_{i=1}^{N} U_i + U_0 \right), \tag{5.9}$$

where U_0 is the elastic energy of the substrate and U_i is the potential energy of the ith actuator. Consider the dependence of the ECE on the parameters of the POE controlled by an actuator with a piezoelectric effect. We assume that the control voltage is applied to the kth actuator.

The energy spent by the actuators to deform the substrate is

$$U_0 = P/2 \left[\Delta l_k + \sum_{i=1, i \neq k}^{N} q_i \Delta l_i \right], \tag{5.10}$$

where Δl_i is the elongation of the ith actuator. In the actuator with a control signal applied, the stored energy is

$$U_k = -P\Delta l_k + \sigma l D_3 E_3, \tag{5.11}$$

where E_3 is the electric field strength applied to the actuator; D_3 is the electric induction arising in the actuator material; l is the length; and σ is the cross sectional area of the actuator. The energy stored by the actuator due to its deformation by the substrate is

$$U_i = P/2q_i \Delta l_i \quad (i = 1, \ldots, N; i \neq k). \tag{5.12}$$

Therefore, the ECE is determined by the expression

$$\eta = [f_k(\rho_k, \theta_k) - G \sum_{i=1, i\neq k} q_i^2]/\{k_{31}^{-2} G^{-1} [f_k(\rho_k, \theta_k) + G]^2 - f_k(\rho_k, \theta_k) - G\}, \tag{5.13}$$

where $k_{31} = d_{31}(s_{11}\varepsilon_{33})^{-1/2}$ is the piezomechanical coupling coefficient; d_{31} is the piezoelectric modulus; s_{11} is the elastic compliance; and ε_{33} is the permittivity of the actuator material. In the case of magnetostrictive actuators, formula (5.13) instead of the piezomechanical coupling coefficient includes the magnetomechanical coupling coefficient.

Thus, the ECE depends on the mutual arrangement and the number of actuators and two dimensionless parameters—relative rigidity G and coefficient k_{31}. Figure 5.1 shows the dependence $\eta(G)$ for the actuators of the 19-element POE, located at the nodes of a hexagonal grid with a spacing $b = 0.45R$ (the coefficient $k_{31} = 0.3$). The dependences have a maximum corresponding to the optimal (in terms of efficiency of the substrate deformation) relative rigidity G. We draw readers' attention to the smallness of the ECE (typically, $\eta = 0.002$). To this end, the selection of the optimal arrangement of the actuators and coordination of their mechanical properties with the substrate in order to ensure a minimum energy loss is fundamentally important in the development and fabrication of adaptive POEs.

Figure 5.1a shows the dependences of the maximum ECE on the distance between the actuators for a hexagonal arrangement of 19 actuators. Figure 5.1b demonstrates the relative rigidity corresponding to the maximum ECE ($k_{31} = 0.3$) distances, and so it is impossible to choose the distance b between the actuators and relative rigidity G so as to simultaneously provide the optimal control energy conversion into the strain energy of the substrate for all the actuators. However, dependences presented make it possible to determine b and G, corresponding to a maximum total.

As an example Fig. 5.1 shows the dependences $\eta(k_{31})$ for the actuators of the 19-element adaptive EPOEP with a hexagonal arrangement at $b = 0.3R$ and $G = 0.2$. The dependences of the ECE on k_{31} have the same character for the actuators located at different distances from the center: the ECE increases with increasing k_{31}. In the case of the peripheral actuators, with increasing k_{31} the ECE changes faster than for the actuators located closer to the center; however, this difference is insignificant.

The longitudinal deformation of the piezoceramic actuator is caused by transverse or longitudinal piezoelectric effects. The relationship between electrical and mechanical variables in the case of the transverse piezoelectric effect has the form

$$S_1 = s_{11}T_1 + d_{31}E_3,$$
$$D_3 = d_{31}T_1 + \varepsilon_{33}E_3, \tag{5.14}$$

where S_1 is the longitudinal deformation of the actuator; T_1 is the mechanical strain in its cross section; E_3 is the electric field strength applied to the actuators; D_3 is the electric induction arising in the actuator material; d_{31} is the piezoelectric modulus; s_{11} is the elastic compliance; and ε_{33} is the permittivity of the actuator material.

Electrical and mechanical variables can be related to the elongation Δl of the actuator, the force F and the control voltage U by the following expressions:

$$S_1 = \Delta l/l, \quad T_1 = F/\sigma, \quad E_3 = U/d, \tag{5.15}$$

where l is the length, σ is the cross-sectional area, and d is the transverse dimension of the actuator. Then, the elongation of the actuator has the form

$$\Delta l = s_{11}l\sigma^{-1}F + d_{31}ld^{-1}U, \tag{5.16}$$

where the coefficient before the loading has the sense of the inverse rigidity of the actuator, $s_{11}l\sigma^{-1} = K_a^{-1}$, and the second term of the sum corresponds to the elongation of the actuator in free state. Introducing appropriate notations, we rewrite in a more compact form:

$$\Delta l = K_a^{-1}F + \Delta l_0, \tag{5.17}$$

where $\Delta l_0 = d_{31}ld^{-1}U$.

Rod- and spring-type magnetomechanical devices of precision displacement, implementing magnetostrictive Joule and Wiedemann effects, can be used as actuators.

For a rod-type actuator the relationship between the magnetomechanical variables has the form

$$B_3 = \mu_{33}H_3 + d_{33}T_3, \quad S_3 = d_{33}H_3 + s_{33}T_3, \tag{5.18}$$

where B_3 and H_3 are the induction and strength of the exciting magnetic field; T_3 and S_3 are the stress tensor components of deformation resulting from the implementation of the conditions of the magnetostrictive Joule effect; μ_{33} is the absolute permeability of the magnetostrictive material of the actuator; d_{33} is the magnetostrictive modulus; and s_{33} is the mechanical compliance of the magnetostrictive material of the actuator.

The force F and elongation Δl of the actuator have the form

$$\Delta l = S_3 l, \quad F = T_3\sigma, \tag{5.19}$$

where l is the length of the actuator and σ is its cross section; when a current I flows through the coil of the actuator with the number of windings N, the axial magnetic field strength is $H_3 = NI/l$. Thus, the elongation of the actuator to which an external force F is applied has the form

$$\Delta l = K_a^{-1}F + \Delta l_0, \tag{5.20}$$

where $K_a^{-1} = s_{33} l \sigma^{-1}$ is the rigidity of the actuator and $\Delta l_0 = d_{33} NI$ is the elongation of the actuator in free state.

In the case of a spring-type actuator the combination of axial and circular magnetic fields, N_{ax} and N_{circ}, and inductions, B_{ax} and B_{circ}, in the cross section of the spiral of the actuator produce a torsional strain ϕ_{tr} that the spring transforms into linear displacements Δl:

$$\Delta l = 2\pi R^3 nF/[\cos\alpha(B\sin^2\alpha + C\cos^2\alpha)], \tag{5.21}$$

where F is the force acquired by the actuator; n is number of windings of the actuator; R is the radius of the spring of the actuator; α is the helix angle of the actuator; and C and B are the rigidity of the cross section of the actuator spiral to torsional and bending stresses, i.e., the actuator rigidity

$$K_a = \cos\alpha(B\sin^2\alpha + C\cos^2\alpha)/2\pi R^3 n.$$

The constitutive equations for the magnetostrictive material under the conditions of the Wiedemann effect fulfillment allow one to obtain an expression for the running angle of the twist of the cross section of the actuator spiral φ through a magnetic field H_{ax}:

$$\varphi = 2 \cdot 2^{1/2} \kappa_{ii} h_{ijk} H_{\mathrm{ax}} G^{-1} r^{-1} F(r/l_0), \tag{5.22}$$

where κ_{ii} and h_{ijk} are the tensor components of the magnetic susceptibility and the magnetoelastic coupling; G is the shear modulus; r is the radius of the cross section of the actuator spiral; $F(r/l_0) \sim 1$ is a dimensionless function describing the dynamics of the magnetization of the actuator spiral by an alternating magnetic field in a constant field; and l_0 is the skin-layer depth.

The relationship between the angle φ and the displacement Δl_0 induced by the actuator in the absence of external mechanical stresses has the form

$$\Delta l_0 = 2\pi R^2 n\varphi \cos^{-2}\alpha. \tag{5.23}$$

Expressions allow one to establish a similar to (5.23) dependence, where K_a is determined from (A2.4) and

$$\Delta l_0 = 4\pi R^2 n 2^{1/2} \kappa_{ii} h_{ijk} NI/\cos^2\alpha Grl. \tag{5.24}$$

These dependences well describe the operation of magnetostrictive actuators of rod and spring types and are widely used in practice.

Chapter 6
Iterative Synthesis Algorithm of the Shape of the Adaptive POE Reflecting Surface

In this chapter, we consider the possibility of using linear iterative synthesis algorithm of the shape of a flexible adaptive POE under conditions of nonlinearity of hysteretic electromechanical properties of the individual actuators and their connection through a flexible substrate. We present the results of the experimental realization of the iterative algorithm and demonstrate its fast convergence.

Creation of adaptive systems involves the use of various adaptive POEs whose reflecting surface shape is controlled by elastic deformations of the substrate by system of discrete actuators made of piezo- or magnetostrictive materials. Actuators have nonlinear hysteretic transfer characteristics and are connected with each other with the help of an elastic substrate, thereby resulting in non-orthogonality of the response function of the adaptive POE. Therefore, the control can be implemented by either nonlinear algorithms or linearized iterative procedures. We present the experimental substantiation of possible formation of a reflecting POE surface with an error traditional for optics (~0.1 μm) by using a linear iterative algorithm.

It was previously shown that the synthesized shape of the reflecting POE surface, described by the function $W(r, \phi)$, is a superposition of the surface response function with the weighting coefficients determined by the electrical and mechanical properties of the actuators and control signals applied to these actuators:

$$W(r,\varphi) = \sum_{i=1}^{N} F_i U_i f_i(r,\varphi), \tag{6.1}$$

where $W(r, \varphi)$ is the position of the points r and φ of the reflecting surface with respect to some support surface (including the initial surface of A3); $f_i(r, \varphi)$ is the response function of the adaptive POE to the control signal (1 V) applied to the ith actuator; F_i is the weighting coefficient determined by the electromechanical properties of the ith actuator; and U_i is the voltage of the control signal applied to the ith actuator. In the case of the linear dependence of $W(r, \varphi)$ on the control signals and linear independence of the response function, expression (6.1) is used

V.V. Apollonov, *High-Power Optics*, Springer Series in Optical Sciences 192,
DOI 10.1007/978-3-319-10753-0_6

to calculate U_i, defining the function $W(r, \varphi)$. For $W(r, \varphi)$, specified at the nodes of the grid covering the POE aperture, U_i can be found from the system of linear equations:

$$\sum U_i f_{ij} = \Delta W_j, \qquad (6.2)$$

where f_{ij} is the value of the response function at the jth point of the aperture to the action ith actuator; $\Delta W_j = (W_j - W_j^T)$ is the deflection of the POE surface at the jth point of the aperture; W_j is the required deflection relative to the reference plane at the jth point of the surface; W_j^T is the current deflection relative to the reference plane at the jth point of the surface.

Control signals from (6.2) provide the formation of the adaptive POE surface with high accuracy due to the nonlinearity and hysteresis characteristics of the actuators. However, using the linear iterative procedure the elementary step of which is determined by formula (6.2), we can in principle increase the accuracy of the synthesis of the reflecting surface shape, provided that this procedure ensures a quick convergence and stability. The theoretical study of the convergence and stability of the proposed iterative procedure involves considerable difficulties, and so in this paper it is investigated experimentally.

The iterative algorithm is experimentally tested using the automated system, the optical path of which is a laser Fizeau interferometer. The interferogram was recorded with a TV camera and displayed on the black and white television monitor. The device digitizes the TV signal and ensures the storage of one image, its output to the screen and input of the signals into a computer. The interferogram is processed in the fringes of finite width by an algorithm that uses the dispersion relations for the logarithm of the analytical signal. The program calculates the deflections ΔW_j required for the synthesis of a given surface shape of the POE at fixed points of the aperture and by solving the system of equations (6.2) computes the control actions U_i. The object of the study was a flat metal POE with a diameter of the reflecting surface $D = 12$ cm, controlled by a system of 19 piezoceramic actuators which are located at the nodes in a hexagonal grid with a spacing of 21 mm. The surface of the adaptive POE had a square subaperture with a side of 6 cm. The surface was shaped by using seven central actuators. The control voltage was applied by the switching procedure on the control panel while the power supply to compensate for the capacitive load was connected via a matched inductance.

As a synthesized surface we used a sphere with a 2μm deflection within the sub-aperture. Formation was performed in two stages. At the first stage we measured the response function f_{ij} per unit effect of each actuator at six points of the surface. In this case the control voltage was applied to the ith actuator and the discrete response function at given points of the surface was calculated as the difference between the POE surface deflections before and after the application of the control voltage. The resulting response function was normalized to the voltage applied to the actuator. Deflection of the surface was calculated relative to the point of attachment of the central actuator, the position of which was fixed

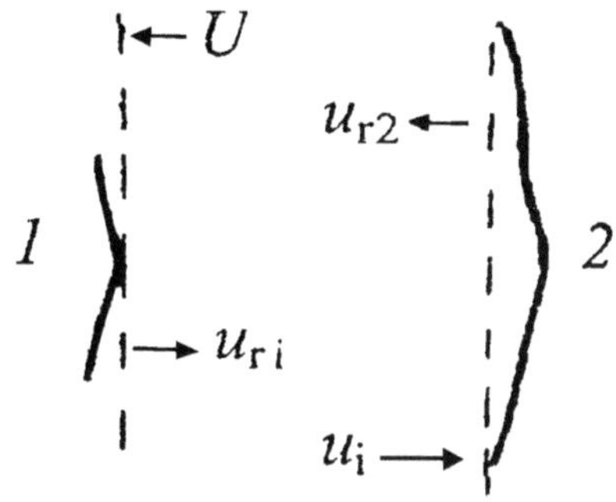

Fig. 6.1 Scheme of resonator. *1*—output POE, *2*—deaf POE, *U*—given field, u_i—direct field, u_r—reflected from POE fields

due to stabilization of the control voltage applied to it. The result of the measurements is the matrix of the coefficients f_{ij}. The second step was to form a given spherical surface. In this case we determined the shape of the current surface and calculated the residual ΔW_j relative to a given surface at the points for which the matrix is generated. Then, by solving the system of equations (6.2) we calculated the required stress increments on the actuators, which were added to the stresses on the actuators. Thereafter, the resulting surface was controlled and, if necessary, the procedure was repeated. To exclude the saturation of the actuator the control voltage was limited to ±200 V. The sequence of synthesis of a given spherical surface is shown in Fig. 6.1. The initial POE surface of random shape after four iterations was used to produce a spherical surface with an rms error of 0.25 λ (λ = 0.63 μm). Further iterations did not give any significant decrease in the rms error.

Chapter 7
Synthesis of Laser Resonator Configurations as an Inverse Problem in Optics

We have developed an approach within the framework of which the problem of searching and selecting the configuration of the laser cavity was formulated as an inverse optical problem: the resonator configuration, i.e., complex reflection coefficients of the resonator adaptive POEs are found from the specified amplitude and/or phase distribution of the field in the plane of the output aperture. We obtained closed integral equations to determine complex reflection coefficients and, in some cases, found exact analytical solutions. The problem of development of the deformation algorithm of adaptive POEs to control the laser power was formulated and solved as an inverse problem of synthesis of the geometric configuration of the resonator for a given phase distribution of the output radiation. The results of numerical studies of the output radiation power of controlled resonators and axial radiation intensity in the far zone were presented. It was shown that in the selection of one- or two-mirror resonator configurations it is expedient to proceed primarily from the comparative level of complexity of production of deformable adaptive POEs and control systems.

It is well known that the characteristics of the laser radiation are considerably dependent on the configuration of the optical resonator, i.e., the geometric shape and the type of apodization functions of resonator adaptive POEs. Search and selection of the optimal resonator configuration is a necessary stage in the development of laser systems. A number of studies is known, which are devoted to the study of the properties of cavities with non-spherical POEs. The development of the methods and tools of adaptive optics has revived interest in resonators with deformable POEs and other correctors of the phase field distribution. In the adaptive control regime resonators with adaptive POEs allow one to partially compensate for intracavity phase field distortions and bring the characteristics of laser radiation into proximity with those in the absence of the field distortion. The tools of adaptive optics are also promising for solving the problem of searchless formation of laser radiation with desired characteristics, which vary in time in accordance with a given program. In this regard, the development of laser resonators relies on the approach in which the configuration of the resonator is determined by the characteristics of the output radiation as the solution to the inverse optical

V.V. Apollonov, *High-Power Optics*, Springer Series in Optical Sciences 192,
DOI 10.1007/978-3-319-10753-0_7

problem. In this paper we consider the problem of searching for the resonator configuration, which allows one to generate output radiation with a given amplitude and/or phase distribution.

7.1 Resonator with a Given Field

The inverse optical problem of searching for the cavity configuration by the given field distribution in the plane of the output aperture of the resonator consists in the following (Fig. 7.1): complex reflection coefficients A_1 and A_2 are found for resonator adaptive POEs 1 and 2, respectively, which provide, in a plane containing the output aperture of POE 1, the creation of the resonance field U, specified up to a constant factor and propagating in the direction from the resonator.

Let the field U be defined on the entire infinite plane containing the aperture of POE 1 and POE 2 infinitely. We find the relation between $A_{1,2}$ and the field U. The complex reflection coefficient A_2 is equal to the ratio of the field u_{r2} reflected from POE 2 to the field u_i incident on the same POE.

The field u_{r2} is calculated by using the function U with a reversible ray path taken into account [1], the field u_i—taking into account the reflection from POE 1. In the Fresnel–Kirchhoff approximation

$$u_{r2} = U_{r2}\exp(-ikL), \quad u_i = U_i\exp(ikL),$$

where

$$U_{r2}(\vec{r}_1) = \int U(\vec{r}_2)K_{12}^{*}d\vec{r}_2; \quad U_i(\vec{r}_1) = \int U(\vec{r}_2)A_1(\vec{r}_2)K_{21}d\vec{r}_2; \tag{7.1}$$

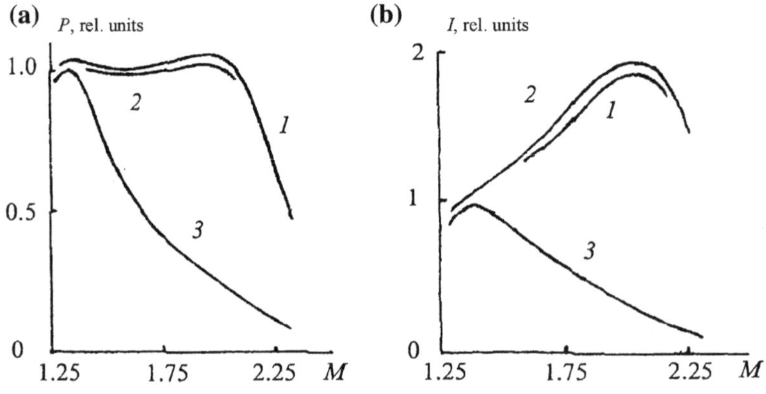

Fig. 7.1 The far-field output radiation power (**a**) and axial power density as functions of the geometrical increase factor M of the reference resonator (**b**). The radiation power at a given M was found from the condition of its maximum versus the amplitude of the POE deformation. The results of calculations for the reference unstable resonator (*curve 3* for NM2 = 12, non-active lose 0.12, and Mth = 2.5) and controlled resonator with one (*curve 1*) and with two adaptive POEs (*curve 2*)

k is the wave number; L is the length of the resonator; $\vec{r}_{12}$ is the radius vector of a point in the plane of the POE aperture in a cylindrical coordinate system; m, $n = 1, 2$; and $K_{mn} = K(\vec{r}_n, \vec{r}_m) = k\{\exp[ik(\vec{r}_n - \vec{r}_m)^2]/2L\}/2i\pi L$ is the kernel of the extension integral (the Green function) [1]. Thus, if the desired, physically realizable reflection coefficient A_2 exists, then

$$A_2(\vec{r}) = \gamma U_{r2}(\vec{r})/U_i(\vec{r}). \tag{7.2}$$

The constant factor γ includes the coefficient exp($-2ikL$) and is determined taking into account the physical limitations $|A_{1,2}| \leq 1$; below we assume that $\max|A_{1,2}| = 1$. The quantity γ contains an undefined phase factor, this uncertainty is removed when additional limitations on the form of the functions arg $A_{1,2}$ are taken into account.

Equation (7.2) defines the desired configuration of the resonator at given U and A_1. Similarly to expression (7.2), the relation for the reflection coefficient A_1 as a function of U and A_2 has the form:

$$A_1(\vec{r})P_1(\vec{r}) = \gamma U_{r1}(\vec{r})/U(\vec{r}), \tag{7.3}$$

where $P_1(\vec{r}) = 1$ within the aperture of POE 1 and $P_1(\vec{r}) = 0$ outside it;

$$U_{r1}(\vec{r}_1)A_2^{-1}(\vec{r}) = \int U_{r2}(\vec{r}_2)A_2^{-1}(\vec{r}_2)K_{12}^{*}d\vec{r}_2. \tag{7.4}$$

The right-hand side of (7.3) must vanish when the point $\vec{r}$ is outside the aperture of POE 1. This condition is equivalent to the equation with respect to A_2:

$$U_{r2}(\vec{r}_1)A_2^{-1}(\vec{r}) = \int U_{r2}(\vec{r}_2)A_2^{-1}(\vec{r}_2)K_{32}^{*}P_1(\vec{r}_3)K_{31}d\vec{r}_2d\vec{r}_3. \tag{7.5}$$

Equation (7.5) is obtained by multiplying expression (7.3) by $1 - P_1(\vec{r})$ with the help of the following integral transformations of the resultant expression. If the aperture of POE 1 is infinite, then condition (7.5) becomes an identity and is excluded from consideration. Thus, formulas (7.3) and (7.4) with account for (7.5) allow one to determine the reflection coefficient of POE 1 at given U and A_2.

It should be noted that when the radiation passes through POE 1, the field distribution of the transition radiation is $u_e \sim UF(A_1)$, where F is the known function that takes into account the impact of the output POE 1 on radiation passing through it. If the field u_e is specified, then the above relations can be used with the equation substituting u by $u_e/F(A_1)$. In this case, however, relation (7.3) requires further investigation because it becomes an equation of relativity.

The field U is realizable not only by a couple of interconnected functions A_1 and A_2 and allows one, at fixed U, to optimize the resonator in mode transparency, selectivity and range of resonant frequencies. The equation for determining the mode characteristics of the resonator with a given field is similar to the Fox—Li equation and obtained taking into account relations (7.2) and (7.3) from the condition of the field reproducibility during the round-trip of radiation in the resonator volume

$$\gamma_q V_q(\vec{r}_1) = \int V_q(\vec{r}_2) G_{21} d\vec{r}_2, \tag{7.6}$$

where

$$G_{21} = G(\vec{r}_2, \vec{r}_1) = \gamma A_1(\vec{r}_2) \int K_{23}[U_{r2}(\vec{r}_3)/U_i(\vec{r}_3)] K_{31} d\vec{r}_3 \tag{7.7}$$

at a given A_1 and

$$G_{21} = G(\vec{r}_2, \vec{r}_1) = \gamma [P_1(\vec{r}_2) U_{r1}(\vec{r}_2)/U(\vec{r}_2)] \times \int K_{23} A_2(\vec{r}_3) K_{31} d\vec{r}_3 \tag{7.8}$$

at a given A_2. In the integral (7.6), the transverse field distribution in the wave/mode with $q = 0, 1, \ldots$ incident on POE 1 is ($|\gamma_q| \geq |\gamma_{q+1}|$) (the range of eigenvalues). The field obviously satisfies (7.6), i.e., there is a number s, for which $V_s = U$. The number is determined by studying the entire range of eigenvalues of (7.6). The cavity configuration is optimized in A_1 in formulas (7.6) and (7.7) or in A_2 in formulas (7.6) with the requirements to the range $\{q\}$ and other limitations taken into account.

7.2 Phase Problem for a Resonator with Two Adaptive POEs

In the general case, the complex reflection coefficient obtained from expressions (7.2) or (7.3) for POE 2 or 1, respectively, is implemented by deformation and apodization of adaptive POEs. However, the currently developed tools of adaptive optics directly affect only the phase distribution of the field. Furthermore, the radiation from the resonator is usually coupled out by POE 1, whereas POE 2 is not apodized. With this in mind, of interest is the reverse optical phase problem aimed at determining the shape of adaptive POEs, ensuring the generation of the field U propagating from the cavity with a given distribution in the plane of the output aperture at a specified apodizing function for the output POE 1 in the absence of apodization of POE 2.

We obtain equations for determining the complex reflection coefficients A_1 and A_2. To this end, we present them in the form $A_j = |A_j| \exp[i \arg(A_j)]$, $j = 1, 2$, where $|A_1| \leq 1$ is the given apodization function of the output POE resonator 1 (for an infinite POE $|A_2| = 1$) and arg A_j are the desired parts of the reflection coefficients. To find the equation for A_j, it is sufficient to express arg A_j through the right-hand sides of (7.2) and (7.3). Thus, we obtain the following system of equations for A_2 and A_1:

$$A_2(\vec{r}) = \exp\{i \arg[\gamma U_{r2}(\vec{r})/U_i(\vec{r})]\}, \tag{7.9}$$

$$A_1(\vec{r}) = P_1(\vec{r})|A_1(\vec{r})| \exp\{i \arg[\gamma U_{r1}(\vec{r})/U(\vec{r})]\}, \tag{7.10}$$

where the functions $U_{r2}(\vec{r})$, $U_i(\vec{r})$, and $U_{r1}(\vec{r})$ are defined by (7.1) and (7.4) and expressed in terms of $A_{1,2}$ and U, whereas the function $P_1(\vec{r})$ in (7.10) takes into account the limitation (7.5) on the class of functions A_2. It should be noted that the complex reflection coefficient A_2 calculated by formula (7.9) [formula (7.10) is used to calculate the coefficient A_1] at given A_1 (or A_2) minimizes the mean square of the difference modulus between the given field U and the field calculated for the round-trip of radiation in the cavity: output aperture → POE 1 → POE 2 → output aperture (or, in the opposite direction: output aperture → POE 2 → POE 1 → output aperture). This extremality property is easy to prove by the method of variational calculus in the study of the formation of the rms error of the field as a functional of arg A_2 (or arg A_1).

The system of (7.9) and (7.10) together with (7.6) allows one to explore all the main characteristics of the resonator, resulting from deformation of adaptive POEs for a given distribution of the output radiation. Numerical solution of system (7.9), (7.10) can be found by iteration. This method is mathematically equivalent to the computational algorithms used to study the extracavity generation of a specified intensity distribution: In this case, specified is the intensity distribution $|U_{r2}|^2$ on POE 2, which is to be formed as a result of the passage along the path: output aperture → POE 1 → POE 2 → output aperture. Convergence of the iterative algorithm is probably due to the above extremality properties of relations (7.9) and (7.10).

In some cases, system (7.9), (7.10) has an explicit analytical solution. Suppose, for example, that both resonators of adaptive POEs are infinite and POE 1 is unapodized. From the analysis of these relationships it is easy to establish that the functions

$$A_2(\vec{r}) = U_{r2}(\vec{r})/U^*_{r2}(\vec{r}), \quad A_1(\vec{r}) = U^*(\vec{r})/U(\vec{r}) \tag{7.11}$$

are the solution to the inverse phase problem (in this case $|A| = 1$). The solution to (7.11) is also valid for resonators with finite dimensions of POE 1 if outside the aperture of this POE the specified field is $U = 0$.

Consider in more detail the case of $U = U^*$, when the given field is homogeneous in phase. Then, according to (7.11) the reflecting surface of POE 1 is flat and

$$A_2(\vec{r}) = \exp[i\alpha_2(\vec{r})], \quad \alpha_2(\vec{r}_1) = 2\arg\left[\int U(\vec{r}_2)K^*_{12}d\vec{r}_2\right], \tag{7.12}$$

where $\alpha_2(\vec{r}) = 2k\delta_2(\vec{r})$ is the control phase effect, and $\delta_2(\vec{r})$ is the deviation of the surface shape of POE 2 from the flat one. Thus, in this version the formation of the amplitude field distribution in a homogeneous phase distribution is achieved by deformation of a totally reflecting POE in the case of fixed plane geometry of the output POE. For example, the field $U(\vec{r}) = 1$ within the aperture of a square output POE with the side $2a$, and $U(\vec{r}) = 0$ outside the aperture; $\vec{r} = (x, y)$ and $|x/a| \leq 1$, $|y/a| \leq 1$. From (7.12) it follows that

$$\alpha_2(\vec{r}) = \beta(x) + \beta(y),$$

where

$$\beta(x) = 2\arg\{erf[(i\pi N)^{1/2}(1 - x/a)] + erf[(i\pi N)^{1/2}(1 + x/a)]\}; \quad (7.13)$$

and $N = ka^2/2\pi L$. When $N \gg 1$, function (7.13) oscillates around zero with an amplitude of oscillation $\sim(1 + \pi N^{1/2})^{-1}$. To implement the phase distribution $\alpha_2(\vec{r})$ requires a POE with the number of degrees of freedom of no less than 8 N.

7.3 Resonator with an Adaptive and Apodizing POE

When radiation is coupled out through a semitransparent window 1, it is technically challenging to dynamically control its shape; therefore, of interest is the inverse problem: to determine the shape of the unapodized adaptive POE 2 and the apodization function of semitransparent windows 1 having a fixed form, which provide the generation of a given field in the plane of the output aperture of the cavity.

We obtain equations for determining A_2 and $|A_1|$, by considering the function U, which describes the field distribution in the plane of the output aperture of the resonator, and the function arg A_1, which is associated with the shape of POE 1, to be given. Because POE 2 is not apodized, the complex reflection coefficient A_2 satisfies the same relation (7.9) as that in case of the phase problem discussed above. Expression (7.9) includes the function U_i, which, in accordance with (7.1), depends on $|A_1|$, and so one needs the second equation to determine $|A_1|$. We obtain it from (7.3) by extracting the moduli of left- and right-hand sides of this relation. Thus, we arrive at the following system of equations with respect to A_2 and $|A_1|$:

$$A_2(\vec{r})\exp\{i\arg[\gamma U_{r2}(\vec{r})/U_i(\vec{r})]\}, \quad (7.14)$$

$$|A_1(\vec{r})| = P_1(\vec{r})|\gamma||U_{r1}(\vec{r})/U(\vec{r})|. \quad (7.15)$$

The solution of (7.14) and (7.15) is based on relations (7.1) and (7.4) defining U_{r2}, U_i, and U_{r1}. System (7.14), (7.15) is easily reduced to a nonlinear integral equation with respect to A_2 or $|A_1|$ by substituting (7.15) into (7.14) or (7.14) into (7.15), respectively. Exact analytical solution of the equations obtained is not possible to find; therefore, the study of system (7.14), (7.15) requires the use of numerical methods.

7.4 Resonator with an Adaptive POE to Generate the Field with a Given Phase or Amplitude Distribution

For a field to be generated with a given phase or amplitude distribution, the resonator should have one adaptive POE. Consider the problem of generation of a field with a predetermined phase distribution arg $U(\vec{r})$, when POE 2 is deformable, and the complex reflection coefficient A_1 is specified. The function $|U|$ of the amplitude distribution of the generated field is not specified; it must be determined simultaneously with the coefficient A_2.

We obtain relations for determining A_2 and $|U|$. We use expression (7.2), which includes both unknown functions. Using the equality $|A_2| = 1$ and equating to unity the modulus of the right-hand side of (7.2), after integral transformations we obtain a closed equation with respect to the amplitude distribution of the output radiation and the modulus of the eigenvalue:

$$|\gamma||U(\vec{r}_2)| = \left| \int |U_i(\vec{r}_1)| \exp\{i \arg[U_{r2}(\vec{r}_1)]\} K_{12} d\vec{r}_1 \right|, \tag{7.16}$$

where the functions $U_i(\vec{r})$ and $U_{r2}(\vec{r})$ are defined by (7.2) and depend on $|U|$ and A_1. The solution of the nonlinear (7.16) with respect to $|U|$ of is a set of eigenfunctions with the corresponding range of eigenvalues. The desired reflection coefficients A_2 are calculated with the help of formula (7.2) or (7.9) by using the substitution $U = |U| \exp[i(\arg U)]$. Numerical solution of (7.16) can be found by iteration.

For the output radiation with a given amplitude distribution $|U|$ to be generated by the adaptive POE 2, we calculate the coefficient A_2 after finding the phase distribution of the output radiation. The equation for determining it [as (7.16)] is derived from (7.2):

$$\arg[U(\vec{r}_2)] = \arg\{ \int |U_i(\vec{r}_1)| \exp\{i \arg[U_{r2}(\vec{r}_1)]\} K_{12} d\vec{r}_1 \}. \tag{7.17}$$

Once arg U is found from (7.17), formulas (7.2) or (7.9) can be used to calculate the reflection coefficient A_2.

7.5 Inverse Problems in the Approximation of the Perturbation Theory

Based on the equations obtained, it is difficult to take into account the limitations which can be imposed, for example, on the mode transparency and selectivity of the resonator being developed. Therefore, of interest is the further development of ways of analytical description of the characteristics of the resonator as a function of the specified field.

We consider the problem of synthesis of the resonator configuration in the approximation of the perturbation theory. We represent the complex reflection coefficients in the form $A_j = A_{js} A_{jc}$, $j = 1, 2$. Here, the coefficient A_{js} corresponds to the reference cavity, i.e., the cavity in the absence of the effects controlling the form of the reflection coefficient; the coefficient $A_{jc} = \exp i\,\alpha_j$ is the desired part of the reflection coefficients; α_j is the control action (in general, a complex function); $\mathrm{Re}\alpha_j(\vec{r}) = 2k\delta_j(\vec{r})$ is the phase control action; $\delta_j(\vec{r})$ is the deviation of the POE surface from the reference one, corresponding to a cavity in the absence of control actions; and Im α_j is related to the controlled function of the POE apodization. In the absence of control actions ($\alpha_j(\vec{r}) = 0$) radiation generated in the plane of the output aperture of the cavity is characterized, as is known, by a set of modes $\{v_q\}$ with the corresponding range of eigenvalues $\{\lambda_q\}$ ($q = 0, 1, \ldots$). Thus, for a reference cavity, the equation

$$A_{2s}(\vec{r}) = \lambda_q v_{r2}^{(q)}(\vec{r})/v_i^{(q)}(\vec{r}), \tag{7.18}$$

which is similar to (7.2), is valid, where

$$v_i^{(q)}(\vec{r}_1) = \int v_q(\vec{r}_2) A_{1s}(\vec{r}_2) K_{12} d\vec{r}_2.$$

We represent the specified field in the form $U(\vec{r}) = v_q(\vec{r}) \exp[i\varphi(\vec{r})]$ and restrict our analysis by the variant, when the complex reflection coefficient from the output POEW 1 is specified. Assuming that the amplitudes $|\alpha_{1,2}|$, $|\varphi| \ll 1$ in (7.2) and (7.18), in the first approximation we find

$$\begin{aligned} \alpha_2(\vec{r}_2) = C + &\int \varphi(\vec{r}_1) v_q(\vec{r}_1) K_{21}^* d\vec{r}_1 / v_{r2}^{(q)}(\vec{r}_2) \\ &- \int [\varphi(\vec{r}_1) + \alpha_1(\vec{r}_1)][v_q(\vec{r}_1) A_{1s}(\vec{r}_1) K_{12}] d\vec{r}_1 / v_i^{(q)}(\vec{r}_2), \end{aligned} \tag{7.19}$$

where C is a complex constant; Im C is found from the condition $\max|A_{2s}(\vec{r}) \exp[i\alpha_2(\vec{r})]| = 1$ and Re C—from additional requirements to the resonance radiation wavelength, the phase amplitude Re α_2, etc. The range of the eigenvalues $\{\gamma_n\}$ of a synthesized resonator is calculated by the known formulas of the perturbation theory in the appendices to the optical resonators:

$$\gamma_n = \lambda_n \exp i[\alpha_1^{(n)} + \alpha_2^{(n)}], \quad n = 0, 1, \ldots, \tag{7.20}$$

where

$$\alpha_1^{(n)} = \int \alpha_1 v_n^2 A_{1s} d\vec{r} \Big/ \int v_n^2 A_{1s} d\vec{r};$$

$$\alpha_2^{(n)} = \int \alpha_2 \{v_{r2}^{(n)}\}^2 A_{2s}^{-1} d\vec{r} \Big/ \int \{v_{r2}^{(n)}\}^2 A_{2s}^{-1} d\vec{r}.$$

Thus, relation (7.19) defines the complex action α_2 controlling the resonator configuration at which the distribution v_q of the field emerging from the reference cavity is transformed into $U = v_q \exp i\varphi$. The position of the mode with the U distribution in the range of eigenvalues is found from the analysis of (7.20). The field distribution in the other resonator modes can be calculated by using (7.19) in the approximation of the perturbation theory. In accordance with (7.19) and (7.20), any mode of the reference cavity can be transformed into a mode with the required distribution of U; however, it should be borne in mind that the solution to (7.19) and (7.20) depending on the specific form of the functions U, φ, v_q, $v_{r2}^{(q)}$, A_{1s}, $v_i^{(q)}$ does not always satisfy the conditions of applicability of the perturbation theory.

Consider in more detail the case when the reference resonator is made of infinite spherical reflecting surfaces. We restrict our research to the transformation of the fundamental mode ($q = n = 0$) of the reference cavity. The field distribution in this mode in the absence of control actions is described by Hermite or Laguerre functions. Using the explicit expressions for the field of the fundamental mode and complex reflection coefficients of spherical surfaces of the reference cavity, relation (7.19) for the control actions will have the form

$$\begin{aligned}\alpha_2(\vec{\rho}) = C + iN \int \Big\{ &v_1\varphi(\vec{\rho}_1)\exp\left[-iNv_1(\vec{\rho}_1 - \vec{\rho}/v_1)^2\right] \\ &+ v_2^{-1}\left[\alpha_1(\vec{\rho}) + \varphi(\vec{\rho}_1)\right]\exp\left[i\pi N(\vec{\rho}_1 - \vec{\rho}v_2)^2/v_2\right]\Big\} d\vec{\rho}_1,\end{aligned} \tag{7.21}$$

where v_1 (or v_2) is the eigenvalue of the integral equation, which is part of the Fox—Li equations and corresponds to the round trip of radiation from POE 2 to POE 1 (or from POE 1 to POE 2); $\vec{\rho} = \vec{r}/a$; $\vec{r}$ is the dimensional radius vector of the POE aperture; a is the characteristic size of the resonator adaptive POEs; and $N = ka^2/2nL$ is the Fresnel number.

The approximate relation (7.21) can also be used for resonators with Gaussian amplitude reflection coefficients of the POE. In this case, $v_{1,2}$ depend not only on the curvature of the surface, but also on their effective size. Complex quantities $v_{1,2}$ satisfy the inequality $\mathrm{Im}v_{1,2} \le 0$, which is an important condition for convergence of the integrals from (7.21).

For optical resonators, the Fresnel number is generally $N \gg 1$, which allows one (in studying the inverse resonator problems) to use the geometrical optics approximation. Let us compare the solution of (7.21), which describes the transformation of the fundamental mode of a reference cavity in the first-order approximation of the perturbation theory, with the solution in the geometrical optics approximation. To this end, we substitute the formula $U(\vec{r}) = v_0(\vec{r})\exp[(i\varphi(\vec{r})]$ into expression (7.2), and assume, as in the derivation of (7.21), that the reference cavity has a spherical POE. The integrand $\exp[i\varphi(\vec{r})]$ in the numerator of formula (7.2) is expanded in a Maclaurin series around the point $\vec{r}/v_1$. We also expand the integrand $\exp i[(\alpha_1(\vec{r}) + \varphi(\vec{r})]$ in the denominator of (7.2) in the same series with the center $\vec{r}v_2$. After integration and summation of series arising in the numerator and denominator of (7.2), for the control actions up to the accuracy of the terms of the order $1/N$ inclusively, we obtain the expression

$$\begin{aligned}\alpha_2(\vec{\rho}) = \{&C + \varphi(\vec{\rho}/v_1) - \varphi(\vec{\rho}v_2) - \alpha_1(\vec{\rho}v_2) \\ &- i\Delta_\rho[v_1\varphi(\vec{\rho}/v_1) + \varphi(\vec{\rho}v_2)/v_2 + \alpha_1(\vec{\rho}v_2)/v_2]/4\pi N\} \\ &+ [v_1[\nabla_\rho\varphi(\vec{\rho}/v_1)]^2 + [\nabla_\rho\varphi(\vec{\rho}v_2) + \alpha_1(\vec{\rho}v_2)]^2/v_2]/4\pi N.\end{aligned} \tag{7.22}$$

where ∇_ρ is the gradient operator and $\Delta_\rho = \nabla_\rho^2 A_p = V_p$ is the Laplace operator. If integration of expression (7.21) is performed with accuracy up to the terms ~$1/N$, the corresponding formula for $\alpha_2(\vec{\rho})$ will contain the terms in braces in (7.22).

It follows from the analysis of (7.22) that the quadratic terms can be neglected when the amplitude $\sigma_\alpha, \sigma_\varphi$ of the variation functions α_1, φ satisfy the inequalities

$$\sigma_\alpha \ll \min\{N_\alpha^{1/2};\, 1 + \pi N_\alpha\},$$
$$\sigma_\varphi \ll \{N_\varphi^{1/2};\, 1 + \pi N_\varphi/(1 + l_\varphi/a|1 - \lambda_0|)\}.$$

Here $N_{\alpha,\varphi} = N(l_{\alpha,\varphi}/a)^2$; l_α and l_φ are the characteristic scales of variation functions α_1, φ. The first of two quantities under the sign 'min' was obtained from the condition that the value of the quadratic term is much less than unity, and the second—from the condition of smallness of the quadratic term in comparison with the linear one. Contribution of the quadratic term is particularly important in the case of a reference cavity with a nearly plane-parallel geometry. Thus, for $N \gg 1$ $N \gg 1$, $N_{\$\$}$ $\gg 1$ and $N_\varphi/(1 + l_\varphi/a|1 - \lambda_0|) \gg 1$ the applicability of the first-order approximation of the perturbation theory in the amplitude in the variation functions α_1, φ is much wider than that assumed in the derivation of (7.21). In the limit $N \to \infty$ relation (7.22) takes the form

$$\alpha_2(\vec{\rho}) = C + \varphi(\vec{\rho}v_1) - \varphi(\vec{\rho}v_2) - \alpha_1(\vec{\rho}v_2). \tag{7.23}$$

Formula (7.23) is a zero geometrical optics approximation and valid for resonators with reference geometry that is different from the plane-parallel geometry.

Relations (7.19)–(7.21) in the first approximation of the perturbation theory solve the problem of the generation of a given field by the cavity, describe the main characteristics of the resonator and can be considered initial phase in the solution of the phase problem, the problems of generation of output radiation with the desired amplitude and phase distributions. For expressions (7.19)–(7.21) to be used, we should introduce appropriate restrictions on the nature of functions $\alpha_{1,2}, \varphi$. Note also that expression (7.22) is a generalization of a similar relation obtained in relation to unstable resonators. This expression can be used in studies of reconfigurable resonators in the case of both unstable and stable geometries of the reference cavity.

7.6 Resonator with One Adaptive POE for a Laser with a Controlled Output Power

Based on expressions (7.20)–(7.23) we studied the algorithms of the control of the geometric configuration of the resonator which ensure the laser power control. Consider resonators with confocal unstable reference geometry of the positive branch with the adaptive POE 2 and radiation output at the edge of the spherical POE 1. When controlling the laser radiation power its divergence should be close to the diffraction limit, and so the function of the phase distribution of output radiation $\mathrm{Re}\,\varphi(\vec{\rho}) = \varphi_R(\vec{\rho})$ satisfies the equation $\varphi_R(\vec{\rho}) = 0$ for points $\vec{\rho}$ lying outside the aperture of the output POE 1. For points $\vec{\rho}$ lying within the aperture of

POE 1 the function $\varphi_R(\vec{\rho})$ is non-zero. In controlling the shape of POE 2 the form of the function $\phi_R(\vec{\rho})$ within the aperture of POE 1 and the transparency of the resonator change, thereby ensuring the control of the output radiation power. We assume below that the control actions are quasi-stationary; therefore, the explicit time dependence on the resonator characteristics is neglected.

Taking into account the above remarks about the form of the function $\phi_R(\vec{\rho})$ of the phase distribution of the field in the output aperture plane of the POE, we consider the design of the resonator for controlling the laser radiation power as a problem of the resonator synthesis for a given phase distribution of the output radiation. Consider relation (7.22). Excluding the diffraction effects at the edges of the output POE 1 the eigenvalues for the reference cavity are $\nu_1 = 1$ and $\nu_2 = 1/M$ (M is the geometric magnification). Using in (7.22) only the terms linear in φ, α_2, setting $\alpha_1(\vec{\rho}) = 0$ and using in (7.22) real and imaginary parts, we obtain the system of equations

$$\begin{aligned} \mathrm{Im}\alpha_2(\vec{\rho}) &= 0 \\ &= \mathrm{Im}C - \varphi_1(\vec{\rho}) + \varphi_1(\vec{\rho}/M) \\ &\quad - \Delta_\rho[\varphi_R(\vec{\rho}) + M\varphi_R(\vec{\rho}/M)]/4\pi NM^2, \end{aligned} \tag{7.24}$$

$$\begin{aligned} \mathrm{Re}\alpha_2(\vec{\rho}) &= \alpha_2(\vec{\rho}) \\ &= \mathrm{Re}C + \varphi_R(\vec{\rho}) - \varphi_R(\vec{\rho}/M) \\ &\quad - \Delta_\rho[\varphi_1(\vec{\rho}) + M\varphi_1(\vec{\rho}/M)]/4\pi NM^2, \end{aligned} \tag{7.25}$$

where $\varphi_R = \mathrm{Re}\,\varphi$ is the given the phase distribution of the output radiation; $\varphi_1 = -\mathrm{Im}\,\varphi$ is the logarithm of the amplitude distribution of the generated field; $N = ka_1^2/2\pi L$ is the Fresnel number; a_1 is the radius of the output POE; and $\vec{\rho} = \vec{r}/Ma_1$. Equation (7.24) is obtained by considering the fact that mirror 2 is not apodized ($\mathrm{Im}\alpha_2(\vec{\rho}) = 0$). From (7.24) up to a constant, we find that

$$\begin{aligned} \varphi_1(\vec{\rho}) = \sum_0^\infty &\{(M+1)[\Delta_\rho\varphi_R(\vec{\rho})\big|_{\rho=0}] \\ &- M^{(2m+1)}\Delta_\rho\{\varphi_R(\vec{\rho}/M^m) + M\varphi_R[\vec{\rho}/M^{(m+1)}]\}\}/4\pi NM^3. \end{aligned} \tag{7.26}$$

From (7.25) with accuracy up to terms ~$1/N$ inclusive we obtain the relation of the corresponding phase control actions:

$$\alpha_2(\vec{\rho}) = \varphi_R(\vec{\rho}) - \varphi_R(\vec{\rho}/M). \tag{7.27}$$

In the derivation of (7.27) it is assumed that the radiation wavelength of the reference and synthesized resonators is the same. Eigenvalues for the resonator modes are calculated according to (7.26) and (7.27). With accuracy up to terms ~$1/N$ inclusive,

$$\gamma_n = \{M\exp[(M+1)\Delta_\rho\varphi_R(\vec{\rho})\big|_{\rho=0}/4\pi NM^3]\}^{-(n+1)},\ n = 0, 1, \ldots \tag{7.28}$$

Thus, relations (7.26)–(7.28) describe the main characteristics of a controlled resonator with a totally reflecting adaptive POE in a reference unstable confocal resonator of the positive branch.

To find out the requirements to the POEs and justify the scheme of distribution of forces deforming the mirror, we specify the form of the function $\varphi_R(\vec{\rho})$. Most smooth is a bell-shaped function $\varphi_R(\vec{\rho})$ within the aperture of mirror 1. Accordingly, as shown in (7.27), in the case of a 2D resonator the reflecting surface of the POE should have a crater-like relief centered at the top, and in the case of a 1D (strip) cavity a relief with two convexes (or concaves) should be present in the cross section of the POE. One of the least difficult technical scheme of loading of an adaptive POE in a 2D resonator is realized by using a ring drive. In this case, the POE edges and the top, which coincides with the center of the ring drive, must be fixed. It follows from the analysis of relation (7.28) that the transparency $\delta = 1 - |\gamma|^2$ of a controlled resonator is determined by the transparency of local near-axial cavity formed by POE 1 and the deformed portion of POE 2. If the edges of the crater are displaced out of (from) the cavity, the field amplitude $|U| \sim \exp[\varphi_1(\vec{\rho})]$ becomes maximal (minimal) near the axis of the resonator, and the transparency of the controlled resonator for the fundamental mode becomes smaller (larger) than the transparency of the reference cavity.

The results obtained allow us to formulate a number of consequences, which are important for the rational organization of the resonator configuration control.

1. Control algorithm (7.27) allows generalization, similar to that obtained for resonators with two POEs, namely:

$$\alpha_2(\vec{\rho}) = \Psi(\mu\vec{\rho}) - \Psi(\vec{\rho}/M), \tag{7.29}$$

   where $\Psi(\vec{\rho})$ is a smooth bell-shaped function; and $\mu > 0, \mu \neq 1/M$. The analysis of expression (7.29) together with the relation of geometric optics for the field at the cavity output shows that if the spatial scale l_Ψ of the change in the function $\Psi(\vec{\rho})$ is so small that $l_\Psi/2Ma_1 \leq \min\{\mu M; 1/\mu M\}$, and at the same time is large enough so that the conditions for the applicability of the zero geometrical optics approximation are met, the divergence of the output radiation is close to the diffraction limit and the algorithm of the POE shaping corresponding to expression (7.29) can be used to control the power of the laser radiation. For μ it is preferable to use a number of values $\mu = M^m$, where $m = 0, 1, \pm 2, 3, \ldots$
2. In controlling the transparency of the resonator by the deformation of the output POE 1 the POE control algorithm is similar to algorithm (7.29):

$$\alpha_1(\vec{\rho}) = \Psi(\mu M\vec{\rho}) - \Psi(\vec{\rho}).$$

3. The field at the resonator output in the zero geometrical optics approximation is linear in control action functions. Therefore, the deformation function of each POE of the resonator with controlled transparency can be represented as the sum of the deformation functions previously defined for control circuits with one and two adaptive POEs.

These generalizations provide an additional degree of freedom in selecting the deformation functions of the POEs, which makes it possible to optimize their loading scheme and simplify the design of the adaptive POE and the control system.

Further research on the effectiveness of the control algorithm (7.27), (7.29) was performed using numerical methods. The dependence of the gain on the radiation intensity corresponded to a homogeneously broadened line of the working transition; the infinitely thin reinforcing layer was located near the totally reflecting POE. The desired shape of the POE was formed by a linear load uniformly distributed along three lines parallel to the fixed edges of the strip POE. The appropriate relief was in the form of two waves symmetrical with respect to the POE axis. The loads were matched so that the points at its edges and on its axis parallel to the edges were immobile. Spacing between the load lines closest to the POE axis was assumed to equal to its half-width. The shape of the reflecting surface was calculated by using the thin plate approximation.

Figure 7.1 (curve 1) shows typical results of the calculations of the far-field output radiation power and axial power density as functions of the geometrical increase factor M of the reference resonator. The radiation power at a given M was found from the condition of its maximum versus the amplitude of the POE deformation. For comparison, Fig. 7.1 shows the results of calculations for the reference unstable resonator (curves 3) and controlled resonator (curve 2) with two adaptive POEs. In the case of a resonator with two adaptive POEs we assumed that their relief is formed by the load uniformly distributed over the area of each of them, and the edges are fixed. The POE deformation was calculated by using the thin plate approximation. Loadings on the mirrors were matched so that $\alpha_2(0) = -\alpha_1(0)$; in this case, the control algorithm $\alpha_2(x) = -\alpha_1(x/M)$ was implemented.

From Fig. 7.1a it is clear that at $M \approx 1.3 - 2.1$ the output radiation power P of the controlled resonators in question remains virtually constant and close to the maximum power generated by a reference laser at $M \approx 1.32$. The levels of the output power for a resonator with one and two adaptive POEs differ by no more than 10 %. The axial radiation intensity I in the far zone (Fig. 7.1b) for both controlled resonators is maximum at $M \approx 2$ and 1.8–1.9 times higher than the maximum intensity corresponding to the reference cavity at $M \approx 1.4$. The POE deformation amplitude for the controlled resonator with one adaptive POE in this case was 2–5 times less than for a resonator with two adaptive POEs.

The results obtained indicate that the radiation power control schemes using a resonator with one and two adaptive POE have almost the same efficiency. Therefore, when choosing a particular option, it is advisable in the first place to consider the possibility of the manufacturing technology of POEs and complexity of systems of their shape control.

In summary we can say that we have presented an approach, within which the complex reflection coefficients of the POE resonator, which determined the configuration of the resonator and provide an output radiation with a given amplitude and/or phase distribution, are found as the solution to the inverse optical problem. We have obtained closed integral equations relating the complex reflection coefficients with the characteristics of a given field. We have studied the algorithm of deformation of a totally reflecting POE in a confocal unstable resonator for the laser power control. The problem of developing the algorithm is formulated as a problem of synthesis of a geometric resonator for a given phase distribution of the output radiation. Analytical description of the algorithm is obtained in the approximation of the perturbation theory by using the amplitude of the deformations of a totally reflecting POE. It is shown that an adaptive POE should have a crater-like relief centered at the top. We have proposed the scheme of the POE loading with a load distributed along the circumference at its fixed edges and top. We have presented the results of comparative numerical studies of the output power and the axial intensity of the radiation in the far zone for resonators with one and two adaptive POEs in a reference confocal unstable resonator, which are used for controlling the power of the laser radiation. We have shown that in selecting one-or two-element schemes of the resonator configuration control it is expedient to proceed primarily from the comparative level of complexity of creation of adaptive POEs and control systems.

Reference

1. V.V. Apollonov, A.I. Barchukov, A.M. Prokhorov, in *Proceedings of the First European Conference on 'Lasers and Applications'*, Drezden, GDR, 1972

Chapter 8
High-Power Laser Beam Coupler Using a Phase Diffraction Grating

8.1 Statement of the Problem

An increase in the radiation intensity of laser systems has dramatically complicated the control of the main parameters of the generated light beams. Measurements of the radiation power with total absorption calorimeters allow one to use these devices in a wide range of energies and wavelengths. However, such devices are not transmission-type gauges and, hence, exclude the simultaneous or subsequent use of the laser beam for purposes of the experiment [1, 2]. At a high-power density in the beam the probability of failure of the devices is high. Thus, for spherical calorimeters the fraction of absorbed energy is governed by the size of the input aperture. Even at a relatively small diameter of the optical beam, either the calorimeter dimensions become too large or the beam focusing is required. Although scanning calorimeters are transmission-type devices, thermal deformations of the POE distort the spatial characteristics of the beam. When measuring the divergence of the radiation of high-power laser facilities, the energy density in the focal region of the beam is critically close to the optical breakdown threshold. Therefore, the search for the ways of deflecting and attenuating radiation, while retaining its spectral characteristics, is still urgent.

Consider a diffraction laser beam coupler based on a cooled POE and assess the necessary requirements for the system.

8.2 Diffraction of Light by a Phase Grating

A specular reflecting diffraction grating is the most suitable device for operation at high radiation powers as compared to the amplitude grating. In this regard, we will consider below only a phase diffraction coupler based on a cooled POE.

V.V. Apollonov, *High-Power Optics*, Springer Series in Optical Sciences 192,
DOI 10.1007/978-3-319-10753-0_8

Let the initial wave field be described by the distribution $E_0(x, y, z, t)$ and the z axis be chosen perpendicular to the plane of the diffraction coupler. Assuming that the grating localization region along the z axis δ_z is small, i.e.,

$$\frac{2\pi\lambda\delta_z}{L^2} \ll 1, \tag{8.1}$$

where λ is the wavelength of light and L is the grating period, the wave after the coupler can be represented as

$$E(x, y, z, t) = \tilde{E}_0(x, y, z, t)\exp[i\alpha(x, y, t)], \tag{8.2}$$

where

$$\tilde{E}_0(x, y, z, t) = E_0(x, y, -z, t) \tag{8.3}$$

in the case of the coupler, and

$$\tilde{E}_0(x, y, z, t) = E_0(x, y, z, t) \tag{8.4}$$

in the case of diffraction on a bulk grating.

We represent $\alpha(x, y, t)$ as follows:

$$\alpha(x, y, t) = \alpha_0(x, y, t) + \rho(x, y, t), \tag{8.5}$$

where $\rho(x, y, t)$ at $t = t_0$ is a real periodic (quasi-periodic) function of the coordinates with the periods T_x and T_y. In this case, Im $\alpha_0(x, y, t)$ describes the absorption of light by the diffraction coupler, and Re $\alpha_0(x, y, t)$—the additional aperiodic phase modulation of a light wave caused by the inhomogeneity of the optical thickness of the coupler, the deformation and the roughness of the surface, etc.

For a sinusoidal grating

$$\rho(x, y, t) = C(x, y, t)\sin\left[\frac{2\pi x}{T_x} + \varphi(x, y, t)\right], \tag{8.6}$$

where the possible inhomogeneities of the grating are taken into account by the explicit dependence of the amplitude $C(x, y, t)$ and phase $\varphi(x, y, t)$ on the coordinates. It is assumed that $C(x, y, t)$ and $\varphi(x, y, t)$ are weakly dependent on with the coordinates in comparison with T_x and T_y. In this case,

$$E(x, y, z, t) = \tilde{E}_0(x, y, z, t)\exp\left[i\alpha_0(x, y, t)\right]\exp\left\{iC(x, y, t)\sin\left[\frac{2\pi x}{T_x} + \varphi(x, y, t)\right]\right\} \tag{8.7}$$

or, given the identity

$$\exp(iz\sin t) = \sum_{m=-\infty}^{\infty} \mathrm{Im}(z)\exp(imt) \tag{8.8}$$

where Im(z) is the mth order Bessel function,

$$E(x,y,z,t) = \sum_{m=-\infty}^{\infty} \tilde{E}_0(x,y,z,t)\exp[i\alpha(x,y,t)]\mathrm{Im}[C(x,y,t)]\exp[im\ \varphi(x,y,t)] \exp\left(im\frac{2\pi x}{T_x}\right) \tag{8.9}$$

The diffraction angle of the mth order is

$$\sin\theta_m = m\frac{\lambda}{T_x} \tag{8.10}$$

under normal incidence of the light beam, and

$$\sin\theta_m = \sin\theta_0 + m\frac{\lambda}{T_x} \tag{8.11}$$

if the zero-order beam axis makes an angle θ_0 with the normal of the diffraction grating. For the normals of the partial plane waves in different orders of diffraction it follows from (8.11) that

$$\Psi_m = \frac{\cos\theta_0}{\cos\theta_m}\Psi_0, \tag{8.12}$$

where Ψ_m is the angle between the normal of the plane partial wave and mth order diffracted beam axis under the assumption that Ψ_m is small.

Thus, when

$$\begin{aligned} C(x,y,t) &= C_0 \quad \text{at}\, t = t_0 \\ \varphi(x,y,t) &= \varphi(t) \end{aligned} \tag{8.13}$$

the distributions in different orders differ in:

1. constant factors

$$\mathrm{Im}(C_0)\exp[im\varphi(t)] \tag{8.14}$$

2. the change in the angular scales in accordance with (8.12); and
3. the extent of the change in the distribution over the beam cross section

$$\begin{cases} x_m = \dfrac{\cos\theta_m}{\cos\theta_0}x_0 \\ y_m = y_0 \end{cases} \tag{8.15}$$

Since the differences between the parameters of light beams in different diffraction orders are determined only by the diffraction splitting for a given wavelength and do not depend on the parameters of the initial light beam, the measurements of the beam parameters in the mth order in accordance with the expressions (8.12)–(8.15) may be used to find other parameters, including the zeroth order ones.

Since small quantities of the initial radiation must be deflected with minimal insertion distortions in the main beam to measure the parameters of high-power beams, the first positive and negative diffraction orders are the most suitable for the measurements. In this case, the intensity coupling coefficient takes the form

$$\mathrm{Im}^2(C_0) \cong \frac{C_0^2}{4} \tag{8.16}$$

Effect of polarization of light is insignificant, if enough 'rough' gratings are used, i.e., when

$$T \gg \lambda \tag{8.17}$$

Sufficient accuracy is achieved at $T \cong 5\lambda$; in this case, while $\theta_1 \approx 0.2$ rad.

8.3 Selection of Diffraction Coupler Parameters

For the spatial separation of the diffraction orders, the width of the spatial spectrum of the initial distribution in the light beam must be smaller than the frequency of the first spatial harmonic of the grating. However, since the aperture of the radiation source is always physically bounded, the radiation has a significant spatial spectrum. The presence of 'tails' in the angular distribution of radiation imposes certain limitations on the permissible coupling coefficient and the choice of the grating period, which determines the coupling angle.

For example, let the angular distribution of the radiation intensity by described by a Gaussian function

$$J(\sin u) = J_0 \exp\left(-\frac{\sin u}{\sin u_0}\right)^2, \tag{8.18}$$

where u_0 is the divergence angle at the $1/e^2$ level. For the selected coupling coefficient K and the minimum allowable error η the diffraction angle is given by

$$\sin\theta_1 = \left(\ln\frac{1}{K\eta^2}\right)^{1/2} \sin u_0 \tag{8.19}$$

At $K = 10^{-3}$ and $\eta = 10^{-2}$, $\sin\theta_1 \cong 5\sin u_0$ and for small u_0, $\theta_1 \cong 5u_0$.

To assess the impact of the tooth shape of the grating on the parameters of the diffraction coupler, we assume that

$$|\rho(x, y, t)| \ll 1 \tag{8.20}$$

which follows from the smallness of the required coupling coefficient. With an accuracy up to the second order of smallness

$$E(x,y,z,t) = \tilde{E}_0(x,y,z,t)\exp[i\alpha_0(x,y,t)][1 + i\rho(x,y,t) + \cdots] \quad (8.21)$$

An almost periodic real-valued function $\rho(x,y,t)$ can be represented in the form

$$\rho(x,y,t) = \sum_{n=-\infty}^{\infty}\sum_{k=-\infty}^{\infty} C_{nk}(x,y,t)\exp\left[i2\pi\left(\frac{nx}{T_x} + \frac{ky}{T_y}\right)\right] \quad (8.22)$$

Consequently,

$$E(x,y,z,t) = \tilde{E}_0(x,y,z,t)\exp[i\alpha_0(x,y,t)] \left[1 + i\sum_{n=-\infty}^{\infty}\sum_{k=-\infty}^{\infty} C_{nk}(x,y,t)\exp\left[i2\pi\left(\frac{nx}{T_x} + \frac{ky}{T_y}\right)\right]\right] \quad (8.23)$$

Thus, with an accuracy up to higher order terms the beam intensity in $+1$ and -1 diffraction orders is determined only by the amplitude of the first spatial harmonic of the grating. Thus, for a reflection grating with a rectangular shape of the tooth and off-duty ratio 2, the coupling coefficient for the ± 1-order diffraction has the form

$$k = 16\frac{A^2}{\lambda^2}\cos^2\theta_0, \quad (8.24)$$

where A is the tooth height.

For a symmetric triangular shaped tooth

$$k = \frac{64}{\pi^2}\cdot\frac{A^2}{\lambda^2}\cos^2\theta_0 \quad (8.25)$$

When condition (8.13) is not fulfilled, the diffraction coupler introduces an additional multiplicative error. In this case, the spatial spectrum of the diffracted beams is determined by convolution of the spatial spectrum of the initial image and the spatial spectrum of the function $C(x,y,t)\exp[i\varphi(x,y,t)]$.

If $C(x,y,t)\exp[i\varphi(x,y,t)]$ have a much narrower spectrum, i.e., are slowly changing functions of the coordinate, as compared to the spatial spectrum of the initial beam, the distortions of the angular distribution are insignificant and cannot be taken into account when measuring, for example, the angular divergence. In measuring the intensity distribution over the beam cross section it is necessary to compensate for the inhomogeneity of the coupling coefficient to the grating plane.

Since the POE-based gratings are used to measure the parameters of high-power laser beams, we estimate the effect of thermal deformations on the coupling and distortion coefficients introduced in the spatial distribution of the diffracted beam.

From (8.16) it follows that

$$P_1(T^\circ C) \cong \frac{C_0^2(T^\circ C)}{4}, \quad (8.26)$$

where P_1 is the radiation power in the first diffraction order for the amplitude of the fundamental spatial harmonic of the grating, C_0, at temperature T °C.

For a slow diffraction grating the relative change in the amplitude of the grating during heating at ΔT° has the form:

$$\frac{\Delta C}{C_0} = \alpha \Delta \mathrm{T}^\circ. \tag{8.27}$$

If $\Delta T^\circ = 100^\circ$ and $\alpha = 2 \times 10^{-5}$, then $\Delta C / C = 2 \times 10^{-3}$.

Thus, thermal deformations of the grating introduced small distortions introduced in a sufficiently wide range of temperatures.

8.4 Testing of the Diffraction Coupler

Consider the technological capabilities for producing a specular diffraction coupler at a wavelength $\lambda = 10.6\,\mu\text{m}$ and compare the calculated and experimental values of the coupling coefficients in the first diffraction order for $\lambda = 0.63\,\mu\text{m}$ and $\lambda = 10.6\,\mu\text{m}$.

Most simple for fabrication is the profile of the reflecting surface in the form of regular rectangular lines cut by photolithography, because the cutting of a profile a few hundred angstroms deep is virtually impracticable. We fabricated a copper reflecting grating having a diameter 6 mm and a period 100 lines/mm. The width of light and dark lines was the same. In this case, the condition $\lambda / T < 0.2$ was fulfilled, for which the polarization effects are negligible. An aluminum layer 1,000 Å in thickness was deposited on the copper mirror. Then on this layer we deposited a photoresist layer. After exposure through a photomask, the development in a weak alkaline solution and washing we obtained a sample with a stepped structure. Measurement with the help of the MII-5 interference microscope showed that the step width lies in the range of $50 \pm 2\,\mu\text{m}$ and the edges of the lines are almost vertical. After drying, the existing structure was coated with the reflecting layer of gold. The resulting sample was tested in a beam of 0.63-μm He–Ne laser (LG-126). The intensity in the diffraction orders was measured with the PD-24 K photodiode operating in the short circuit mode, which provides a low dark current ($I_d = 0.58\,\mu\text{A}$). The current corresponding to the incident beam was thus 2.5 mA. A value of K_1 at $\lambda = 0.63\,\mu\text{m}$ corresponds to the amplitude of the phase grating profile $C_0 = 790$ Å. Based on the values of the coupling coefficients, their expected values at $\lambda = 10.6\,\mu\text{m}$ must be, respectively, equal to $K_0 = 0.99$, $K_1 = 2.12 \times 10^{-4}$, and $K_2 = 5.8 \times 10^{-5}$. Experimental verification of the coupling coefficient in the $n = \pm 1$ diffraction order was performed by the same method on a CO_2 laser with a continuous wave power up to 5 kW. The diffraction coupler was installed at an angle $\alpha = 45^\circ$ to the incident beam. The zero order diffraction was directed into a conical absorber. The power coupling into the first order was measured with a thermocouple whose signal was recorded by a loop K-115 oscilloscope. The sensitivity of the scheme was 1.26×10^{-2} W/mm. The measured coupling coefficient was equal to $K_1 = (1.16 \pm 0.12) \times 10^{-4}$. The calculated diffraction angle was equal to 0.13 rad, which made it possible to place the measuring devices at a distance of 1.5 m from the grating. The measured coupling coefficient in the first diffraction order was used to calculate the phase profile amplitude $C_0 \cong 350$ Å. The measured coupling coefficients are in good agreement with the calculated values.

The test results confirmed that the use of metallic phase diffraction gratings as couplers of optical beams is promising. Small distortions introduced into the beam make it possible to fabricate a system of monitoring of the parameters of high-power lasers. Sufficiently large diffraction angles of about $0.1 \div 0.2$ rad provides, at relatively small distances, the separation of diffraction orders even for optical beams of large cross sections.

The possibility of obtaining the coupling coefficients of the grating equal to $K_1 = 10^{-3} \div 10^{-4}$ allows one to use them as the resonator mirrors, providing a simultaneous coupling out of two such beams. In addition, it is possible to use standard measuring equipment designed for small and average powers and optical materials that were not previously applicable because of their lack of thermal stability.

References

1. V.V. Apollonov, E.P. Bochkar, V.Ya. Zaslavskii, V.Yu. Khomich, Preprint FIAN, (1978)
2. V.V. Apollonov, E.P. Bochkar, V.Ya. Zaslavskii, V.Yu. Khomich, QE **6**, 615, (1979)

Chapter 9
Optical Coatings

9.1 Optical Coatings Based on Intermetallic Compounds

This chapter presents the results of experimental studies of intermetallic compound coatings with unique properties important for high-power optics [1–3]. We have investigated the dependence of the absorption coefficients on the chemical and phase composition of vacuum condensates of copper and tin alloys in the tin concentration range from 17 to 60 %, as well as the γ phase of copper–aluminum and copper–gallium systems in bulk and thin-film states. The minimal absorption coefficient equal to 2.1 % is obtained for a Cu_3Sn intermetallide with a tin concentration of 38–40 %. Thus we have shown that the coatings made of the electronic-type intermetallic compounds allow one to produce POE surfaces of high optical quality. The low scattering coefficient of laser radiation, high hardness and oxidation resistance, which significantly increase the lifetime of static and adaptive POEs, bond strength to the substrate, manufacturability are the main advantages of intermetallic coatings, which provided their extensive practical applications.

Recently, interest in intermetallic compounds and their deposition technologies has increased significantly due to the creation of the cooled POEs with capillaries and porous structures [4–6]. However, until now intermetallides have one major drawback—high absorption coefficients, which does not allow them to be used in POEs without highly reflecting coatings.

Optical properties of intermetallic phases are poorly studied; in particular, there is no experimental data on the absorption coefficients, because their theoretical calculation with sufficient accuracy seemed impossible. In our opinion, the study of the optical properties of intermetallic compounds having minimum values of the absorption coefficients without degrading other characteristics is today of great interest as it allows one to give up the operation of deposition of highly reflecting coatings in the manufacture of a wide range of POEs.

This chapter shows the results of experimental studies of the absorption coefficients as a function of the chemical and phase composition of vacuum condensates of copper and tin alloys in the tin concentration range from 17 to 60 % (by

V.V. Apollonov, *High-Power Optics*, Springer Series in Optical Sciences 192,
DOI 10.1007/978-3-319-10753-0_9

weight) and the γ phase of copper–aluminum and copper–gallium systems in bulk (50–100 μm) and thin-film (0.2–0.5 μm) states.

The coatings in question are deposited by evaporation and condensation in vacuum. The composition of the condensates was investigated using a Cameca electron microprobe analyzer in cross-section plane and in the plane of reflection. Optical surface quality of the POE was determined by measuring the scattering coefficient by the screen method at a wavelength $\lambda = 0.63$ μm. In addition the electron-microscopic examination of the structure was carried out by stereoscopic photography of carbon replicas. Absorption coefficients of intermetallic mirror surfaces were measured at a wavelength $\lambda = 10.6$ μm. High optical quality of polished intermetallic surfaces and minor scattering losses allowed us to use, along with the calorimetric method, the method of multiple reflections with accuracy ~5 %.

The results of identification of the chemical composition of the condensates and the component distribution over thickness are presented in the figure, where the stages of the condensate thickness build-up are expressed as a fraction of a full cycle of evaporation. Figure 9.1 shows that in the range of tin concentration from 15 to 45 %, when its content is increased in the evaporated melt, the chemical homogeneity of the coating increases, approaching nearly the composition of the initial charge. These data were used to vary purposely the composition of the intermetallic coatings to be deposited. Table 9.1 presents the absorption coefficients of massive condensates; the data are averaged by the results of the measurements of samples from different lots. The wide range of compositions was obtained by

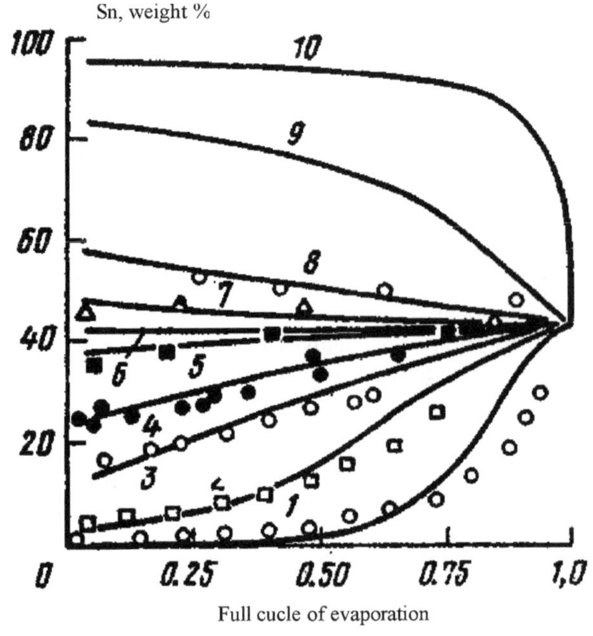

Fig. 9.1 The range of tin vapour concentration (*solid line*-computer data) and condensation phase (experimental dotes) for alloy Cu–Sn with concentration of Sn: 10 (1, ○); 20 (2, □); 30 (3, △); 35 (4, ●); 40 (5, ▪); 42.6 (6); 45 (7, ▲); 50 (8, ○); 70 (9) и 90 % (10); T vap = 1,400 °C

Table 9.1 Coefficients of absorption of intermetallic compounds based mirror surfaces ($\lambda = 10.6$ µ*m*)

Chemical composition of surface, [Sn], % (mass)	Phase composition of surface	Coefficient of absorption, %
0	Cu	1.0
17–19	$\alpha + Cu_{31}Sn_8$	5.7
20	$\alpha + Cu_{31}Sn_8$	6.3
22	$\alpha + Cu_{31}Sn_8$	5.8
28	$\alpha + Cu_{31}Sn_8$	5.3
28–30	$Cu_{31}Sn_8 + \alpha$	5.1
30–32	$Cu_{31}Sn_8 + \alpha$	3.6
32.5	$Cu_{31}Sn_8$	2.8
35	$Cu_{31}Sn_8 + Cu_3Sn$	3.9
36–38	$Cu_3Sn + Cu_{31}Sn_8$	3.6
38–40	Cu_3Sn	2.1
41	$Cu_3Sn + Cu_6Sn_5$	2.7
43	$Cu_3Sn + Cu_6Sn_5$	2.9
44	$Cu_3Sn + Cu_6Sn_5$	2.5
45	$Cu_6Sn_5 + Cu_3Sn$	2.7
50	$Cu_6Sn_5 + Cu_3Sn$	4.0
61	Cu_6Sn_5	3.5

consistently grinding condensate layers parallel to the plane of the substrate and exposing the zones with a monotonically decreasing tin content, as well as by varying the composition of the evaporated metal.

Microstructures of the mirror surfaces in question are a mixture of a α-solid solution of tin in limiting concentration of copper with eutectoid ($\alpha + \delta$) containing 27 % tin, with the intermetallic phase eutectoid δ ($Cu_{31}Sn_8$), intermetallic phases δ and ε (Cu_3Sn), intermetallic phases $\varepsilon + \eta$ (Cu_6Sn_5), and pure phases δ, ε and η (Fig. 9.1).

Based on the results shown in Table 9.1, we concluded that the absorption coefficient is significantly dependent on the phase composition of the surface, but is not a monotonic function of the concentration of tin.

The worst absorption (>5 %) is typical of compounds with a relatively low content of tin, which are presented by α-solid solutions with the face-centered cubic arrangement and by two-phase mixtures ($\alpha +$ $Cu_{31}Sn_3$). With increasing the fraction of the δ-phase in a a two-phase mixture, the absorption coefficient decreases and reaches a minimum (<3 %) for intermetallic surfaces δ, ε-phases and alloys with a predominance of the ε-phase; an increase in the fraction of the η-phase in a two-phase mixture results in a marked increase of absorption.

The minimum absorption coefficient equal to 2.1 % ($\lambda = 10.6$ µm) is obtained in the present work for the Cu_3Sn intermetallic compound with a concentration of tin from 38 to 40 %.

Table 9.2 shows the microhardness of pure metals and intermetallic compounds, arranged in ascending order of the absorption coefficients. From this it follows that in the absorption coefficient the intermetallic phases are comparable to those commonly used in POEs made of such metals as molybdenum, tungsten, nickel, beryllium and aluminum. The microhardness of these intermettalides is considerably higher, which results in a better quality of polishing and smaller coefficients of laser light scattering. Therefore, the total loss on the absorption and scattering of mirror surfaces of intermetallic compounds in some cases is even less than for the highly reflecting metals, not to mention other advantages of these coatings.

The ability of the material to be processed by diamond powder, determined basically by its hardness, affects the value of the absorption coefficient, which depends strongly on the polished mirror surface defects in the form of residual grains of diamond powder. For the dependence of the absorption coefficient on the surface defects of intermetallic coatings to be experimentally determined, we deposited films 200–500 nm in thickness on the substrates of high optical quality. In this case, we avoided possible mechanical damage of the reflecting surface, caused by its polishing.

The measurement results confirm the good polishing ability of the investigated intermetallic compounds. The change in the absorption coefficients for all samples did not exceed tenths of a percent (for example, for Cu_3Sn the difference was 0.1 %).

The presented results of experimental studies of the dependence of the absorption coefficients of optically polished surfaces of condensates and thin films of intermetallic compounds on the chemical and phase composition are of

Table 9.2 Absorption coefficients ($\lambda = 10,6$ mkm) and hardness of metals and intermetallic phases

Исследуемый материал	Коэффициент поглощения, %	Микротвердость, кг/мм^3
Cu	0,97	76
Ag	1,06	25
Mg	1,46	31–48
Al	1,77	19
W	2,02	340
Cu_3Sn	2,1	560
Be	2,56	26
Mo	2,67	160–200
$Cu_{31}Sn_8$	2,83	537
Ni	3,2	180
Cu_9Ga_4	3,3	640
Cu_6Sn_5	3,5	365
Cu_9Al_4	4,5	480
Va	5,57	182
Tl	8,58	106–230

considerable interest both from the point of view of their applicability in the manufacture of POE coatings and understanding the relationship between the composition, structure and optical properties of electronic-type intermetallic phases.

9.2 Optical Dielectric Coatings

Thus, we have conducted a series of studies and developed the technology of manufacturing highly reflective multilayer mirrors. In the experiments, it was found that at a level of rear mirror deformations corresponding to the certain optical power, the combined effects of mechanical stresses and exposure to laser radiation lead to failure. To reduce the stress, a multilayer combined metallic dielectric coating consisting of 3–5 layers was developed (in a traditional coating there are 20–25 layers). To ensure the radiation resistance of coatings use was made of the technology of ion bombardment and implantation. The coatings were tested experimentally by lasers with an output power up to 25 kW. These tests showed high radiation resistance in the high-frequency pulse-periodic regime. Ion treatment significantly improves the coating deposition technology and the optical and operational characteristics, such as radiation resistance, scattering and absorption coefficients. The most important areas of ion treatment when applying optical coatings are as follows:

- ion polishing of wafers in vacuum chamber for coating deposition;
- ion polishing of coatings (ion treatment of the produced film surface);
- ion sub-polishing of coatings (layer-to-layer ion polishing of the film during its deposition);
- ion bombardment of the growing film during its deposition;
- deposition of coatings by ion beam sputtering;
- deposition of coatings by ion beam sputtering with ion bombardment of the growing film;
- deposition of coatings from an ion beam incident on the substrate.

Ion treatment of optical materials for production of mirrors for high-power lasers is an effective tool for advanced optical technology. Ion treatment has already made it possible to elaborate a number of innovative technological processes that can significantly enhance the precision and performance capabilities of high-power laser industry, which is especially important for mirrors of high-frequency pulse-periodic lasers [5, 6]. The value of the scattering coefficient of 0.1–0.2 μm-thick Si films is 0.015–0.02 %. Microphotographs show that the surface of conventional films is significantly nonuniform: against the general background of particles measuring 7–10 nm there is a significant amount of clusters measuring 25–60 nm. After ion polishing the surface of films becomes uniform, the crystallite size being 5 to 7 nm. The coefficient of light scattering after ion polishing with the help of oxygen ions with energy up to 3 keV for 5 min decreased to 0.005–0.008 %. Radiation resistance of optical coatings of the mirrors increases significantly because due to the appropriate choice of the regimes of ion bombardment, the damage threshold of optical coatings increases by 65 %, and the lifetime of the mirror increases by an order of magnitude and reaches the desired value of 3,000 h, which corresponds to today's serviced life of laser diodes used to pump the gain medium. These results are of practical interest in designing optical interference systems with small loss of light, elements of integrated optics and mirrors of high-power high-frequency pulse-periodic laser systems. The developed coatings made it possible to use membrane mirrors in solid-state laser resonators. With these mirrors the feasibility of CW laser radiation *Q*-switching with a deformable rear mirror has been tested experimentally [7].

References

1. V.V. Apollonov, S.I. Derzhavin, V.I. Kislov, V.V. Kuzminov, D.A. Mashkovsky, A.M. Prokhorov, G.I. Babayants, B. Kishmakhov, Sh. Patent 2117371, (1996)
2. V.V. Apollonov, Lasers 1997 (Louisiana, USA, 1997)
3. V.V. Apollonov, V.P. Malyavin, V.N. Feofilaktov, Lasers 1997, (New Orleans, Louisiana, USA, 1997)
4. V.V. Apollonov, Laser Phys. **23**, 1 (2013)
5. V.V. Apollonov, Candidate Dissertation, Moscow, FIAN, 1975
6. V.V. Apollonov, Doctoral Dissertation, Moscow, FIAN, 1982
7. V.V. Apollonov, S.I. Derzhavin, V.I. Kislov, V.V. Kuzminov, D.A. Mashkovsky, A.M. Prokhorov, G.I. Babayants, B. Kishmakhov, Sh., Patent 2117371 (1996)

Chapter 10
POEs Based on Highly Porous Cellular Materials

Study of boiling processes in porous structures is an urgent task because the use of such structures is associated with the possibility of achieving extremely high levels of heat transfer. Analysis of the evaporating liquid inside a heated porous structure [1] shows that a necessary condition for the enhancement of heat transfer is a high hydraulic permeability of the porous structure for a vapor–liquid coolant flow. Most fully these requirements are satisfied by highly porous cellular materials (HPCMs), produced by metallization of polyurethane foam. The structural composition of HPCMs is important for understanding the characteristic features of heat and mass transfer intensification. In the case of HPCMs the cells are virtually identical and represent polyhedrons with cavities having a close-to-prolate ellipsoid of rotation (eccentricity of 1.1–1.3), the major axes of the cells being oriented in the direction of the polyurethane foam (Fig. 10.1a). Stacking of the HPCM cells is close to dense packing—each cell has on average 12 nearest neighbors; the average edge length is 0.4 of the average diameter of the cell. The HPCM cells are interconnected via the 'windows', the maximum size of which is 0.6 of the cell diameter. The HPCM structure is a matrix of the triangular bridges, which form the edge of cell faces; the bridges are secured with their ends in the nodes, i.e., each node has four bridges and belongs to four cells and the bridges belong to three cells. Such a connection provides a high rigidity and hydraulic permeability of HPCMs. Currently these porous structures are made of all the metals and alloys used in technology, the cell diameters range from 0.4 to 5 mm and porosity—from 80 up to 98 %.

The purpose of our study was to determine the heat transfer intensity in HPCMs under conditions of the forced porous filtration flow of a boiling coolant in the porous structure.

Porous boiling of a coolant is as follows: vapor bubbles formed in discrete evaporation centers fill the entire cross section of the pores. As a result of the pore filling, the vapor erupts in the largest interconnected pore channels and flows in the form of separate microjets. Steam microjets are either intercepted by liquid plugs or fill the blind pores. By increasing the heat load intensity the number of steam generation sites and, hence, the amount of steam microjets increase, causing

V.V. Apollonov, *High-Power Optics*, Springer Series in Optical Sciences 192,
DOI 10.1007/978-3-319-10753-0_10

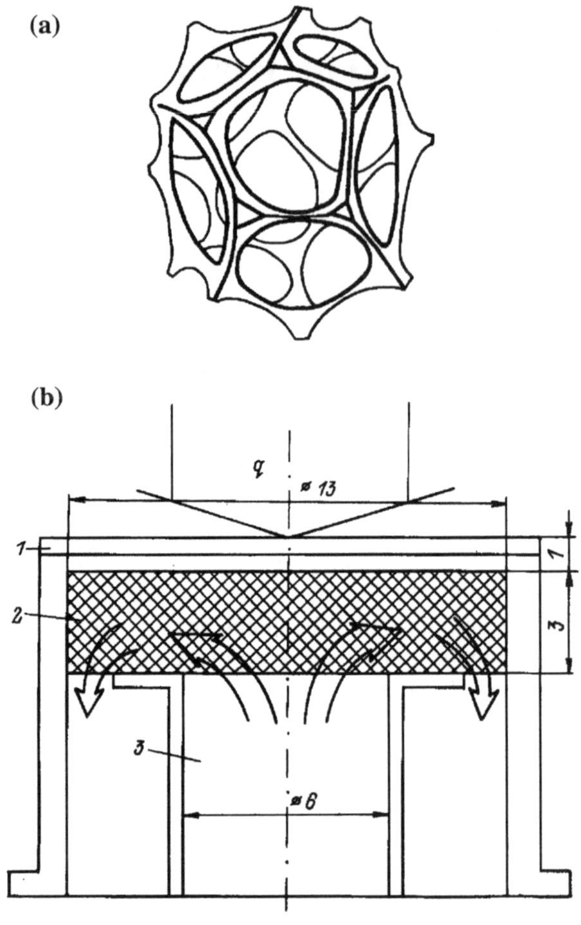

Fig. 10.1 Polyurethane foam based sell (**a**), POE containing a copper HPCM (**b**), *1*—surface of mirror, *2*—heat exchanger, *3*—water pipe

the vapor phase to fill the microfine steam cells. The natural outcome of the process is the heat transfer crisis, i.e., the emergence of a vapor layer separating the coolant flow filtration area and the area of heat removal inside the porous structure. The forced flow of the coolant in the porous structure does not qualitatively change the vaporization situation. It only specifies the direction of the vapor phase and, to some extent, provides the removal of the vapor from the heat zone, thereby shifting the heat transfer crisis to higher thermal loads.

To establish quantitative properties of the heat transfer scheme developed above, we conducted an experimental study of the heat transfer in the POE containing a copper HPCM (Fig. 10.1b) with a porosity $\Pi = 0.81$, average cell diameter of 0.67 mm and skeletal thermal conductivity of $\#_{sk} = 0.3$ W/cm°, soldered inside the body with a 2-layer heat-exchange surface 1 mm in thickness. The heat-exchange surface of the POE was cooled by a coolant jet and had a thermocouple in the center of the zone of heat removal. The heat load, due to the absence of a high-power laser at the moment of the experiment, was simulated by an electron beam. The heat flux density in the experiments was varied by changing the degree of focusing of the electron beam.

An important point in the experiments was to establish the fact of boiling of the coolant in the HPCM of the cooled POE and proper registration of the heat transfer crisis. It is known that the boiling process is accompanied by a characteristic acoustic emission signal, which allows one to control as the beginning of the boiling and so the heat transfer crisis. To this end, the hydraulic path of the POE had a piezoelectric hydrophone installed, whose signal was detected with a spectrum analyzer.

Indeed, the heat transfer crisis, which is a consequence of the transition from the bubble to the film vaporization regime, is accompanied by changes in the acoustic signal power. Figure 10.2a shows the dependence of the acoustic signal power caused by the coolant boiling on the heat load density at a fixed flow rate of the coolant. The quantity q^*, at which the acoustic signal power reached a maximum, was treated as the critical heat flux. The range of the heat loads, smaller than q^*, corresponded to the steady-state heat transfer of the coolant boiling in the porous structure, whereas the range of supercritical heat flows, $q > q^*$, corresponded to a mixed film—bubble boiling of the coolant.

A quantitative measure of the intensity of the heat removal is the heat transfer coefficient, defined as $\alpha = \frac{q}{T_w - T_s}$, where q is the introduced heat flow, T_w is the wall temperature, and T_s is the coolant temperature because at the POE inlet, the coolant (water in this case) had a room temperature. The general dependence of the heat transfer coefficient has the form

$$N_u = C\mathrm{Re}^n,$$

where $N_u = \frac{\alpha \cdot l}{\lambda}$, λ is the thermal conductivity of a liquid, and l is the characteristic size usually taken equal to $l = a/b$. Here α and b are the coefficients of the Darcy–Forchheimer flow equation and $\mathrm{Re} = \frac{\alpha}{b} \cdot \frac{\bar{v}}{\Pi v}$, where $\bar{v}$ is the average pore filtration rate, Π is the porosity, and v is the kinematic viscosity of the coolant. We found the dependence of the heat transfer coefficient on the coolant flow rate (Fig. 10.2b) and heat flux density (Fig. 10.2c). In the range of the heat load densities up to 2 kW/cm^2 we observed a purely convective heat transfer regime; depending on the coolant flow rate α was within 15–25 W/cm$^{2\circ}$. With increasing intensity of thermal loading, the bubble vaporization, intensifying the processes of heat and mass transfer in the boundary layer on the rear side of the heated plate, led to an increase in α up to 30 W/cm$^{2\circ}$, which provides a stable regime of removal of heat loads with a density up to 6 kW/cm^2 (the flow rate of the liquid was almost at the same and equal to 1.1 m/s).

The dependence of α on the coolant filtration rate v is linear at $v > 0.5$. At $v \approx 2$ m/s the value of α reaches 46 W/cm$^{2\circ}$ (Fig. 10.2c).

It should be noted that the dependences $\alpha(v)$ for the convective heat transfer regime and the regime of heat transfer with boiling are similar. On the basis of this assertion we can formulate the following: In our experiments, the coolant boiling in the HPCM was not the main channel of the heat removal. According to

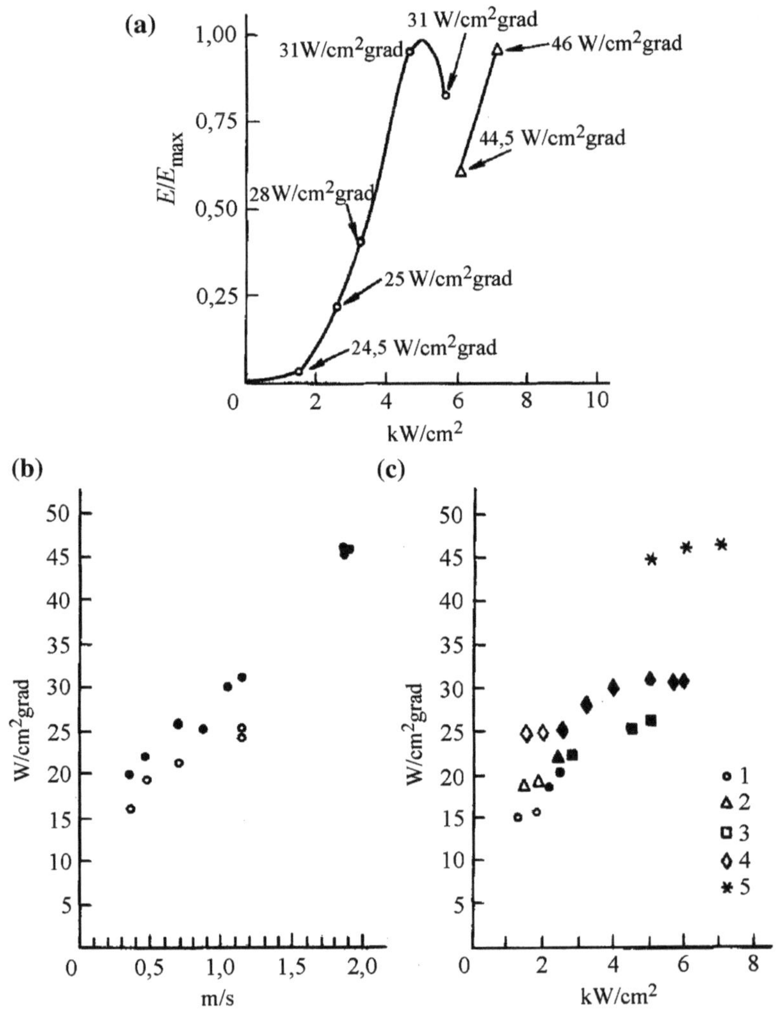

Fig. 10.2 The dependence of the acoustic signal power caused by the coolant boiling on the heat load density at a fixed flow rate of the coolant V = 1.1 m/s (**a**) and the dependence of the heat transfer coefficient on the coolant flow rate (**b**) and on heat flux density (**c**) for different value of V (*1*—0.34, *2*—0.46, *3*—0.68, *4*—1.11, *5*—1.84 m/s)

its basic properties, the heat transfer regime was purely convective and the vapor phase arising in the heating zone of the HPCM was an effective means of coolant mixing, i.e., intensified the process of convective heat transfer in the HPCM. The degree of heat transfer enhancement in this case was more than 1–2 orders of magnitude greater than that on smooth surfaces under forced coolant flow. This allowed us to provide in our experiments a stable regime of heat removal of heat loads with a density up to 7 kW/cm^2 under minor overheating of the heat exchanger for 50 h at moderate coolant flow rates ($v \approx 2$ m/s, $Q = 200$ g/s).

As a result of studies performed, we experimentally identified the main characteristic features of a steady-state regime of evaporation/condensation heat transfer in the POE containing a heat exchanger made of a HPCM. In addition, we proposed a qualitative picture of the process of intensification and experimentally demonstrated the possibility of removal of extremely high density heat fluxes from the POE.

Reference

1. V.V. Apollonov, Doctoral Dissertation, Moscow, FIAN, 1982

Chapter 11
Large POEs Based on Multilayer Honeycomb Structures

Increasing the size of the POEs, while preserving the predetermined level of the optical surface distortion, dramatically increases their mass [1–3]. To reduce the weight of large POEs while maintaining the stiffness of their structures, along with new approaches such as the use of materials with synthesised physical and technical properties promising is also the search for new solutions to the problem. In some cases, the POE weight is reduced by creation of internal voids with relatively large cells. This allows one to decrease the POE weight by 6–7 times for the value of optical surface distortion by its own weight, which is 0.7–0.8 of monolithic mirror distortion. However, it is difficult to create a system of thermal stabilisation without significant loss of rigidity in POEs with large internal voids.

An alternative way to reduce the weight of bulky POEs, as in the case of highly loaded POEs, is the use of highly porous honeycomb materials [4, 5]. We theoretically and experimentally investigated the possibility of creating a lightweight bulky POEs based on multilayer honeycomb structures. Such structures have a relatively small mass at high specific stiffness, good thermal insulation properties and high absorption of elastic vibrations. Multilayer structures also provide the ability to create a highly efficient system of thermal stabilisation.

In the case of axisymmetric thermal loading the problem of thermal distortions of the optical surface of a cooled multilayer honeycomb POE was solved in [1, 2]. In this case, to calculate the temperature fields in a large POE, we considered the problem for a multilayer cylinder whose end and side surfaces were heated, and inside the layers the heat was removed by a coolant. Thermal deformations W^* of the optical surface were determined as the sum of the normal thermal expansion of the POE and its bending

$$W^* = W_{\mathrm{n}}^* + W_{\mathrm{bend}}^*, \tag{11.1}$$

where $W_n^* = \int\limits_0^H \beta(z)T(z,r)dz$ is the normal extension; $\beta(z)$ is the linear expansion coefficient; $T(z, r)$ is the temperature; and H is the POE thickness.

V.V. Apollonov, *High-Power Optics*, Springer Series in Optical Sciences 192,
DOI 10.1007/978-3-319-10753-0_11

Bending was determined from the equation

$$\nabla^4 W_{\text{bend}} = -\nabla^2 \frac{M_T}{D}, \tag{11.2}$$

where

$$M_T = \int_{z_0-H}^{z_0} \frac{E\beta z}{1-\nu} T(z,r)dz \tag{11.3}$$

is the temperature moment;

$$D = \int_{z_0-H}^{z_0} \frac{Ez^2}{1-\nu} dz \tag{11.4}$$

is the bending stiffness; and E is Young's modulus. Poisson's ratio $\tilde{\nu}$ and the position of the neutral surface were determined from the conditions

$$\int_{z_0-H}^{z_0} \frac{E}{1-\nu^2}(\tilde{\nu}-\nu)dz = 0, \quad \int_{z_0-H}^{z_0} Ezdz = 0, \tag{11.5}$$

$$W_{\text{bend}}(r) = C_1 + C_2 \ln r + C_3 r^2 + C_4 r^2 \ln r + \frac{1}{D}\int_0^2 \frac{1}{\rho}\int_0^\rho \xi H(\xi)d\xi d\rho,$$

and the constants C_1, C_2, C_3 and C_4—from the boundary conditions.

Studies showed that for the absorbed heat flux equal to ~10 W cm^{-2} the optical surface distortions of the POE based on multilayer honeycomb invar structures do not exceed 0.7 μm at the POE diameter of 1 m. Constant thermal stabilisation (time needed to reach steady-state operation), which is determined from the solution of the nonstationary problem, for such structures is a few tenths of a second. A peculiar feature of lightweight honeycomb POEs is the fact that a relatively non-rigid filling material may experience a shear strain and transverse compression, significantly affecting the POE operation. In this connection, there appeared a problem of its optimisation, which was considered in the framework of nonlinear programming. The relative displacement of the POE surface under the influence of gravitational, mechanical and thermal loadings was determined by the finite element method.

Figure 11.1 shows the dependence of M^*/M and H^*/H on the allowable distortion W^* of the optical surface of the POE under its own weight. Here M^* and H^* are the weight and thickness of a circular monolithic plate, and M and H are the weight and thickness of the three-layer honeycomb invar structure with a diameter of 2 m. It can be seen that the effectiveness of the multilayer honeycomb structure increases with toughening the requirements for an acceptable distortion of the optical surface. Figure 11.1 also shows that for certain ratios of the structural parameters, the optical surface distortion can be minimised. The example of

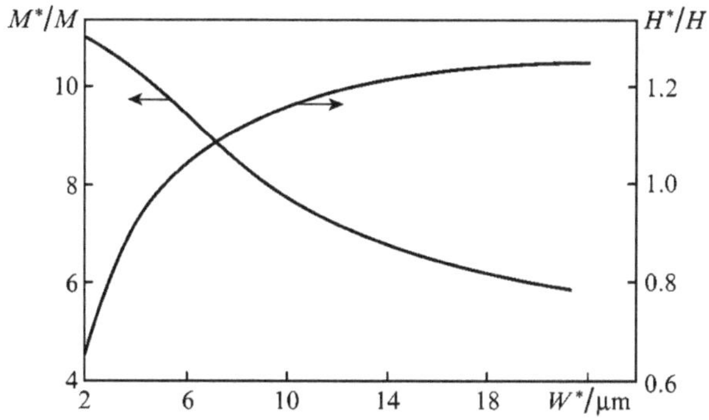

Fig. 11.1 Dependences of the thickness and weight of a bulky honeycomb invar POE 2 m in diameter on distortions of the optical surface

Fig. 11.2 Preform of a large multilayer honeycomb invar POE 1 m in diameter

employment of multilayer honeycomb structures during the manufacture of large POEs 1 m in diameter is shown in Fig. 11.2. Lightweight bulky POEs made of invar are currently used in laser facilities and confirm their high efficiency. This class of POEs is described in detail in [2].

References

1. V.V. Apollonov, Doctoral Dissertation, Moscow, FIAN, 1982
2. V.A. Shmakov, Doctoral Dissertation, Moscow, GPI RAS, 1997
3. V.V. Apollonov, V.A. Shmakov, in *Lasers 1998*, Tucson, Arizona, USA, 1998
4. V.V. Apollonov, G.I. Babayants, P.I. Bystrov, V.F. Goncharov, A.M. Prokhorov, Patent № 244286 (1985)
5. V.V. Apollonov, A.I. Ivlutin, V.N. Motorin, A.M. Prokhorov, V.N. Kharchenko, C.A. Chetkin, Patent № 1278565 (1985)

Chapter 12
Large POEs Based Composite Materials

Progress in this area is largely provided by the rapid development of new technologies and the synthesis of materials with fundamentally new properties. The need for such a development is associated with an ever-expanding range of problems faced by modern science and practice [1–8].

A common disadvantage of large POEs made of glass, glass ceramics, fused quartz and other materials, which are used in optical astronomy and laser technology, is their low thermal conductivity. Such mirrors cannot be used effectively in unilateral heating and technological features of their production do not allow one to significantly reduce weight and ensure effective thermal stabilisation.

Good results in the fabrication of lightweight large POEs have been achieved using composite materials, the methods of their manufacture being well developed. Of greatest interest is the silicon infiltrated carbon-fibre-reinforced silicon carbide composite. The process is based on the deposition of carbon on a free surface during gas phase pyrolysis. Precipitating carbon strengthens frame filaments and combines them into a rigid three-dimensional lattice. The thus obtained porous silica preforms are impregnated with silicon melt in an inert atmosphere. By varying the amount of silicon and impregnation temperature, one can produce samples, significantly different in porosity and phase composition. One can also fabricate virtually carbide porous structures with advanced open porosity, which, except for weight reduction, provides an effective system for thermal stabilisation. Heat treatment removes residual stresses in the composite, increasing its structural stability.

A significant weight reduction of POEs while maintaining their specific rigidity can be achieved also by creating a honeycomb structures. For a honeycomb frame to be manufactured, we used the slip-casting method. Specially prepared slip mass was poured into a mold and polymerised. After removal of the mold the preform was annealed and siliconized.

Figure 12.1 shows a honeycomb frame, produced by the slip casting-method. By joining the resultant honeycomb frame with monolithic plates made of the same material one can form a multilayer honeycomb structure with highly efficient thermal stabilisation.

Figure 12.2 shows a photograph of a lightweight uncooled POE 500 mm in diameter, placed on a polishing/lapping machine. A highly reflecting coating

V.V. Apollonov, *High-Power Optics*, Springer Series in Optical Sciences 192,
DOI 10.1007/978-3-319-10753-0_12

Fig. 12.1 Honeycomb frame of the POE

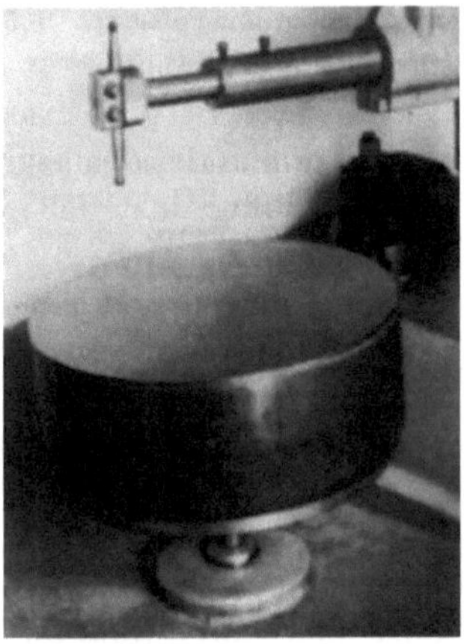

Fig. 12.2 Large POE 500 mm in diameter on a polishing/lapping machine

was deposited on the optical surface of the silicon carbide wafer having a surface roughness of 0.010 μm.

The optical damage threshold of cooled and uncooled POEs based on silicon infiltrated carbon-fibre-reinforced silicon carbide composite was measured experimentally, high power densities being simulated by the electron beam heating facility we developed. The POE was installed in a vacuum chamber and served as an anode of an electro-optical system. The optical damage threshold of the cooled POE 500 mm in diameter was achieved under thermal loading by an electron beam with a power density of ~300 W cm^{-2}, which at characteristic values of the reflection coefficients of the POE materials for laser radiation is equivalent to a power density up to a few tens of kW cm^{-2}. Significant expansion of the range of new materials and the development of modern processing methods and technologies of their connection favours the manufacture of effective large POEs made of C/SiC materials based with record-high thermal stabilisation and high optical performance.

References

1. V.V. Apollono, Doctoral Dissertation, Moscow, FIAN, 1982
2. V.V. Apollono, V.A. Shmakov, *Lasers* (Albuquerque, USA, 2000)
3. V.V. Apollonov, Yu.S. Svirchuk, V.Yu. Guterman, *Lasers* (Albuquerque, USA, 2000)
4. V.V. Apollonov, S.I. Derzhavin, V.V. Kuzminov, Patent №2166820 (2001)
5. V.V. Apollonov, S.I. Derzhavin, V.V. Kuzminov, Patent №2166821 (2001)
6. V.V. Apollonov, S.I. Derzhavin, V.V. Kuzminov, Patent №2166822 (2001)
7. V.V. Apollonov, S.I. Derzhavin, V.V. Kuzminov, Patent №2166823 (2001)
8. V.V. Apollonov, *Lasers* (Tucson, USA, 2001)

Chapter 13
Power Optics and Its New Applications

13.1 Cooling of Laser Diode Assemblies

One of the brightest and most promising implementations of the ideas of power optics is now the introduction of forced heat transfer in high-power semiconductor lasers, which are widely used today to pump solid-state lasers having active elements of different geometry: rods, disks, slabs, fibres [1–4]. Solid-state lasers have the highest efficiency reaching 80 % in some case. Modern manufacturing technologies of semiconductor structures made it possible to significantly increase the laser lifetime (tens of thousands of hours of continuous operation). The variation of the semiconductor material composition can change the wavelength range of radiation from the near-IR to the UV. These lasers are very compact, reliable and easy to operate. The power output can be increased by the simultaneous use of a large number of laser diodes, which are formed in one-dimensional or two-dimensional effectively cooled structures (Fig. 13.1).

Cooled laser diode assemblies possess almost all the remarkable properties of single semiconductor lasers: high intensity, high reliability and long lifetime. These lasers have much smaller weight and size dimensions in comparison with other types of lasers, can easily be fed from independent low-voltage power supplies (solar, nuclear energy) without bulky transformers. Equipment based on laser diode assemblies really becomes a reliable high-performance instrument that can be used in industry, medicine, research and military applications.

The stability of operation of laser diode assemblies and the value of their output are completely determined by the heat transfer efficiency. Laser diode arrays are soldered with a low-temperature solder to the surface of the heat exchanger, which is produced in accordance with high-power optics technology. It should be noted that the levels of heat fluxes which are to be removed from the contact region of the array with the heat exchanger have already approached the characteristic values of the power optics and are equal to several hundreds of W cm^{-2}.

V.V. Apollonov, *High-Power Optics*, Springer Series in Optical Sciences 192,
DOI 10.1007/978-3-319-10753-0_13

Fig. 13.1 Cooled laser diode assembly

To date, there are three main areas of use of cooled high-power laser diode assemblies.

(i) Pumping of solid-state crystal lasers. These lasers, having high-quality output radiation, work as light transformers. The efficiency of such lasers is much higher than the efficiency of solid-state lamp-pumped lasers, because pumping by semiconductor lasers makes it maximal. For kilowatt lasers, the efficiency has already exceeded the value of a few tens of percent. Radiation of semiconductor lasers does not deteriorate the gain medium of solid-state lasers, thereby extending significantly the lifetime of the entire laser system. Weight and size dimensions of such systems are small, which makes them transportable. The main purpose of modern research is to develop and fabricate diode arrays with the highest possible power density. The record of 200 W from a 1-cm-long diode array has been demonstrated recently thanks to the improvement in technology and assembly structure and to a highly efficient heat transfer in the porous structure of the heatsink. The maximum output power from one array, according to calculations, can reach 500 W. This is especially important for further scaling of the power of laser systems and for increasing their efficiency [3–5]. A diode-pumped laser with an output power of 100 kW and higher has been recently developed and fabricated in the U.S.

(ii) Coupling of the diode array radiation into the fibre for puming fibre lasers or for further transport to the processing facility. Radiation losses in the laser diode assembly—optical fibre system are about 15–20 % and mainly determined by the heat transfer, which does not allow one to effectively combine the lasing zones of separate diodes of the array into a minimal spot on the aperture of the fibre with the help of cylindrical optics.

(iii) Development of phase-locked 1D and 2D diode arrays. The output radiation of such lasers is a set of narrowly focused interference peaks (in particular one peak) with a low divergence. Based on theoretical estimates, the output power of such systems can be as high as the power of a conventional out-of-phase diode array. Therefore, such lasers are a source of light with a high efficiency inherent in semiconductor lasers and with a low divergence characteristic of solid-state crystal lasers. At present, such systems are being actively studied. The possibility of scaling the process of phase-locking of arrays and bars of laser diodes has already been shown and high-power laser diode assemblies have been produced.

Thus, the development and fabrication of water-cooled high-power laser diode bars and arrays are promising direction for the implementation of the ideas of power optics in new high-power sources of coherent radiation [4, 5].

13.2 New Generation Power Optics Based on Silicon Carbide

Currently, the development of high-power optics stimulates three trends of efficient use of its technical and technological solutions:

- lightweight, highly stable, large ground- and space-based telescopes for studying the universe and transmitting energy over long distances;
- astronomical optical instruments for remote sensing of the Earth and near space from spacecrafts;
- highly efficient cooled POEs for high-power lasers and laser systems.

All the three trends are based on cutting-edge technologies. The choice of the POE material is a key issue in production of a new generation of optical objects. Thus, a bulky silicon carbide POE has a weight that is 7–10 times lower than that of the POE made of glass ceramics, the best quality in terms of radiation scattering, high thermal stability and a minimum time constant (Fig. 13.2). Comparative evaluation of materials with the help of optical quality criteria developed by us in the early 1970s showed that silicon carbide has a distinct advantage over traditional materials. This conclusion is consistent with more recent conclusions of foreign experts from Germany, France, Japan and China. It is appropriate to note the contribution E.P. Velikhov, who initially supported the creation of the technology of silicon carbide production and the development of large optics.

Obviously, the development of bulky POEs for high-power lasers and transition to a new generation of space-based telescopes is accompanied by the introduction of silicon carbide and related technologies into everyday practice.

In concluding this part of theoretical and experimental works in the field of high power optics we should note one very important point. Effective development of any of the areas of modern cutting-edge technologies, as a rule, gives a result

Fig. 13.2 Silicon carbide POE

not only in the related fields of technological applications, but also in completely different branches of science and technology. Thus, the appearance of one- and two-dimensional cooled laser diode arrays with high radiated power, large astronomical cooled POEs based on silicon carbide and complex composite materials is largely a consequence of the success of power optics [2, 4]—a recognized effective donor for many areas of science and advanced technology of the 21 century. Its successful development continues.

References

1. V.V Apollonov, S.I. Derzhavin, V.A Filonenko, *Lasers* (Tucson, USA, 2001)
2. V.V Apollonov, A.M. Prokhorov, A.H. Guenther, Laser focus. World 1, 101 (2003)
3. V.V Apollonov, in *Proceedings of the Symposium POEM-09*, Wuhan, China, 2009
4. V.V Apollonov, in *Proceedings of HPLS@A-2012*, Istanbul, Turkey, 2012
5. V.V Apollonov, Chin. J. Opt. **6**(1), 1–8 (2013)

Part II
High Repetition Rate Lasers and New Applications

Chapter 14
High Repetition Rate High-Power Wide-Aperture Lasers

14.1 Introduction

A technique for obtaining a repetitively pulsed operating regime in high-power wide-aperture lasers is proposed and experimentally realised. In this regime, the laser emits a train of pulses with a duration of 0.1–1 μs and a pulse repetition rate of several tens of kilohertz. The main properties of the pulsed regime are theoretically analysed and the proposed technique is tested in detail employing a test-bench gas-dynamic laser. The results of the test confirmed the conclusions of the theoretical analysis. The possibility of realising a repetitively pulsed regime in high-power wide-aperture lasers without a reduction in the average output power is experimentally demonstrated.

In our days, interest is increasing in high-power lasers (up to 100 kW) employed in the solution of a variety of research and production problems. The existing sources of high-power radiation operate only in the cw or quasi-continuous low-frequency (below 300 Hz) repetitively pulsed regime with a long pulse duration (tens of microseconds). The development of lasers operating in a high-frequency (tens of kilohertz) pulsed regime with a short pulse duration (hundreds of nanoseconds to few microseconds) or the conversion of the existing lasers to this regime will considerably extend the field of application of high-power lasers, improve the efficiency of their use by factors of several tens, and enable the realisation of qualitatively new effects [1]. For example, to attenuate the plasma screening in the radiation-material interaction, weaken the thermal radiation defocusing in long paths, improve the energy extraction efficiency in wide-aperture lasers, etc.

At a high output power exceeding several kilowatts, however, organising transient lasing modes based on high-frequency resonator modulation runs into several problems, which are caused by wide apertures of resonator elements and accordingly of the laser beam as well as by the high power density.

Presently known devices intended for resonator loss modulation may be conventionally divided into several classes: opto-mechanical, acousto-optic, electro-optical, and self-bleaching. In high-power lasers, only opto-mechanical devices

V.V. Apollonov, *High-Power Optics*, Springer Series in Optical Sciences 192,
DOI 10.1007/978-3-319-10753-0_14

can be used which include transparent or reflective apertures. The remaining modulator types involve transmission optical elements.

In the ten-micrometer range, all optical materials possess a relatively high (up to several percent) absorption coefficient, which is responsible for a significant heat release and, in the long run, a fast degradation of these elements. The use of intracavity disc modulators in high-power industrial lasers is restricted by the output power of several hundred watts: due to the high power density inside the resonator, plasma is produced at the modulator aperture edges to cause modulator degradation or beam screening. In particular, the output power of the CO, laser investigated in [2] (a cw output power of 5 kW) lowered by two orders of magnitude when the laser was converted to the repetitively pulsed regime with the aid of a mechanical tull-aperture modulator. The approach proposed in [3] appears to be more promising—modulating the gain of the active medium rather than the cavity loss. In this work, the gain of the active medium was modulated by imposing a strong external pulsed magnetic field. However, in this case there emerged almost insuperable difficulties related to setup scaling for larger volumes of the active medium as well as to increasing the modulation frequency and the pulse contrast ratio. The authors of [3] were able to raise the modulation frequency to only 10 Hz in a series of only several hundred pulses. This resulted in a reduction in the output power in the repetitively pulsed regime by almost an order of magnitude compared to the cw regime. In the injection of external signal, the methods of modulating the gain of the active medium appear to be the methods of choice [4].

Our work is concerned with a new technique of modulation of the gain of the active medium by radiation self-injection. This technique can be applied to obtain a repetitively pulsed regime in the range of average output power of the order of 100 kW. The aim of our work is to theoretically substantiate and experimentally realise the repetitively pulsed regime of a gas-dynamic CO_2 laser.

14.2 Substantiation of Resonator Design

In lasers with a high average output power, unstable resonator configurations are commonly used because of a large cross section of the active medium. In resonators of this type, externally injected low-power beams may exert a significant effect on the characteristics of output radiation [4, 5].

One way to realise the control regime is the selt-injection of radiation—extraction from the resonator and return of a part of radiation after changing its spatio-temporal characteristics. The transition to the transient lasing mode is effected through the modulation of the self-injecting beam. Earlier, a study was made of laser versions with radiation self-injection into the paraxial resonator region [4]. However analysis showed that the power of the beam injected into the paraxial beam region should be comparable with the output laser power to efficiently control the resonator of a continuously pumped laser, unlike pulsed systems with regenerative amplification [6].

The self-injection of a part of output radiation through the resonator periphery is more efficient: on return to the paraxial resonator region, the injection power significantly rises due to the large number of passages to play the dominant part in the formation of output radiation.

The role of peripheral radiation was first investigated in [4]. In the case of a traditional resonator, the role of waves converging to the resonator axis was found to be insignificant, because their source is a narrow region with a small relative area at edge of the output mirror; accordingly, the power of the control wave injected into the resonator is low. This wave has a large divergence, and only its small part (of the order of $1/Nf$, where $Nf \gg 1$ is the Fresnel number) participates in lasing.

The effect of the injection wave on the resonator characteristics can be enhanced by matching the beam phase with the resonator configuration and increasing the radiation power returned. In this case, the propagation direction and the wavefront curvature of the injection beam should be so matched with the resonator configuration that the injection beam concentrates, after a relatively large number of passages through the resonator, near the optical resonator axis and transforms to a divergent wave that forms the output radiation. The injection beam energy should be high enough to exceed, after its arrival to the resonator axis, the saturation energy of the active medium. The experimental data of the investigation of the effect of this kind of self-injection on lasing in the stationary mode were reported in [5].

The schematic of the setup which realises the repetitively pulsed radiation self-injection is shown in Fig. 14.1. The radiation is extracted from the resonator past the edges of a mirror (2), the mirror coupler (*3*) directs a part of the output beam to

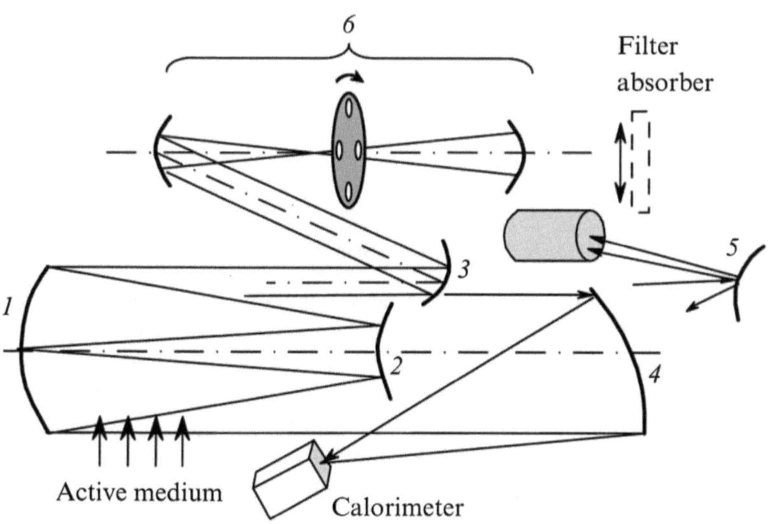

Fig. 14.1 Scheme of the experimental setup: *1, 2* mirrors of the unstable resonator; *3* mirror coupler; *4* rotatory mirror; *5* deflecting mirror; *6* system for the formation of an injected beam

a system intended for the injection beam formation (*6*); the beam is processed in the system (6), where it is modulated in power and acquires the requisite phase distribution, and is returned to the resonator with the aid of the same mirror. The above configuration was realised in a gas-dynamic CO_2 laser with the following parameters; the length of the active medium $L_a = 1.2$ m, the unsaturated gain $g_0 = 0.6\ \text{m}^{-1}$, the time it takes the active medium to transit the resonator $\tau = 0.92 \times 10^{-4}$ s, the relaxation time $\tau_r = 2.76 \times 10^{-4}$ s, the total round-trip time in the resonator $\tau_c = 4.2 \times 10^{-9}$ s, the luminescence lifetime $\tau_{\text{lum}} = 5$ s, the resonator magnification factor $M = 1.45$, and the diameter of output laser aperture $a = 0.08$ m.

The laser resonator is made of two spherical mirrors with rectangular apertures, which provided a geometrical amplification factor of 1.45. The active medium travels across the optical resonator axis. All theoretical and experimental data are given below for a laser with the above parameters. Such a temporal structure of radiation in the case of quazicontinuous mode of operation for pulsed lasers has a name—"rough pulse".

14.3 Theoretical Laser Model and Results of Numerical Analysis

For the initial theoretical treatment of lasing in a gas-dynamic laser with an unstable resonator and transmittance modulation, we will use the modified system of balance equations [7]. In the derivation of equations, the gain of the active medium was spatially averaged over the lasing volume. The gain was assumed to uniformly saturate, decreasing from its peak value (at the point of entry of the active medium into the resonator) to some minimal nonzero value (at the exit from the resonator) with the lateral coordinate. The resultant equations are written in the form coinciding with the form of equations in the case of quasistationary lasing mode:

$$\begin{aligned} \frac{dK}{dt} &= \frac{2K}{\tau}\ln\frac{K_0}{K} - \frac{K[1 + I + (\tau_r/\tau_{\text{lum}})K]}{\tau_r}, \\ \frac{dI}{dt} &= \frac{I}{\tau_c}(K - \delta) + \eta\frac{\tau_r}{\tau_{\text{lum}}\tau_c}K, \end{aligned} \tag{14.1}$$

where $K_0 = 2L_a g_0$ is the averaged unsaturated gain—length product calculated in tracing around the resonator; $K = 2L_a g$ is the averaged saturated gain—length product in tracing around the resonator; $I = J/J_s$; J is the volume-averaged intensity; J_s is the saturation intensity; t is the current time; $\delta = \delta_0(1 + \Delta(\nu))$ are the losses per round trip; $\delta_0 = -\ln(|\gamma^2|)$; $\Delta(\nu)$ is the modulating function; ν is the modulation frequency; and η is the fraction of spontaneous radiation power that remains inside the resonator after tracing around the resonator.

The first of (14.1) is the equation of vibrational kinetics [8] of a preexcited one-component (the lower working levels is not populated) active medium of a gas-dynamic CO_2 laser. The second equation describes the formation of radiation in the propagation through the resonator. The characteristics of active medium and radiation are averaged over the volume, and that is why the equations do not contain directional derivatives and depend only on time.

To determine the conditions ensuring the repetitively pulsed operating regime, the system of equation (14.1) was considered in the perturbation-theory approximation relative to the small parameters

$$\frac{\Delta I}{I_s}, \quad \frac{\Delta K}{\delta_0}, \quad \frac{\Delta \delta}{\delta_0},$$

where ΔI is the amplitude of the deviation of output radiation intensity from the stationary value; $I_s = 2(\tau_r/\tau) \times \ln(K_0/\delta_0) - 1$ is the normalised output radiation intensity for the stationary lasing; ΔK is amplitude of the deviation of the gain—length product from the stationary value; $\Delta\delta$ is the transparency modulation amplitude; $\delta = \delta_0 + \Delta\delta \cos \omega t$; and ω is the circular frequency. The last-named quantity is related to the above-introduced modulation frequency ν in the usual way: $\omega = 2\pi\nu$.

In this approximation,

$$\frac{\Delta I}{\Delta I_s} = \left(\frac{\omega_{\text{res}}\tau}{2}\right)^2 \left\{ \frac{(\omega\tau/2)^2 + 1}{(\omega\tau/2)^2 + [(\omega\tau/2)^2 - (\omega_{\text{res}}\tau/2)^2]^2} \right\}^{1/2}, \qquad (14.2)$$

$$\frac{\Delta I}{\Delta I_s} = \frac{\Delta\delta\tau}{\tau_c} \left\{ \frac{(\omega\tau/2)^2 + 1}{(\omega\tau/2)^2 + [(\omega\tau/2)^2 - (\omega_{\text{res}}\tau/2)^2]^2} \right\}^{1/2}, \qquad (14.3)$$

where $\Delta I_s = (\Delta\delta/\delta_0)(\tau_r/\tau)$ are the quasi-stationary intensity fluctuations (for $\omega \to 0$), and $\omega_{\text{res}} \approx (\mathrm{I}_s\delta_0/\tau_c\tau_r)^{1/2}$ is the resonance circular frequency.

The transition to the repetitively pulsed regime necessitates the fulfilment of two conditions: (i) the transmittance fluctuations should be fast enough, because otherwise the output radiation power will vary in the quasi-stationary manner; (ii) the value of ΔI should be large enough for the radiation intensity to be modulated to a near-zero value.

The former condition is satisfied for $\nu \geq 2/\tau$ and the second for

$$\frac{\Delta I}{I_s} \geq 1 \to \Delta\delta \geq \frac{\tau_c}{\tau} \left\{ \frac{(\omega\tau)^2 + [(\omega\tau)^2 - (\omega_{\text{res}}\tau)^2]^2}{(\omega\tau)^2 + 1} \right\}^{1/2}.$$

For the laser investigated, $\nu_{\text{res}} \approx 100$ kHz, and the repetitively pulsed regime is realised for $\nu > 20$ kHz, $\Delta\delta/\delta_0 > 0.02$.

The resonance field can be represented as a super-position of two waves—the ordinary divergent wave and the convergent one, which transforms to a divergent wave in the incoherent summation in the paraxial resonator region. The transparency δ of the resonator with laser-radiation self-injection, taking into account the diffraction transformation of the convergent wave to the divergent wave in the paraxial resonator region, is defined by the relation-ships

$$\delta = 1 - |\gamma^2|, \quad |\gamma^2| = \frac{1}{M^2} + \frac{s}{|\gamma^{4N}|}, \qquad (14.4)$$

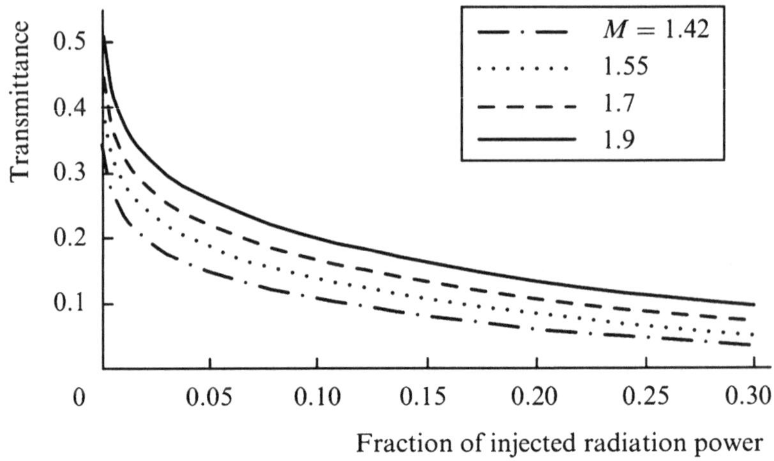

Fig. 14.2 Dependence of laser resonator transmittance on the fraction of radiation power injected into the resonator for different values of resonator magnification factor M

where

$$N \sim \ln\left[\frac{a^2}{\lambda L_r}\left(1-\frac{1}{M}\right)+1\right] \Big/ 2\ln M$$

is the number of transits required of the beam injected into the resonator to find itself in the region of paraxial diffraction transformation; $s = S/\pi a^2$ is the relative injection beam area; S is the injection beam area; and λ is the radiation wavelength.

Figure 14.2 shows the calculated resonator transmittance against beam fraction returned to the resonator. One can see that the modulation amplitude of resonator losses amounts to 30–50 % of the losses of the basis resonator (without self-injection) when the power of the beam returned to the resonator is about 5 % of the output beam power. This loss modulation amplitude is sufficient to ensure the repetitively pulsed operating regime.

To derive qualitative estimates of laser operating modes with sdt-injection, we considered the energy and time characteristics of the laser with the parameters specified above. The system of (14.2) was numerically investigated employing the Runge–Kutta method. Figure 14.3 gives the time dependences of the output power for several values of the modulation frequency and depth; the geometrical resonator amplification factor is $M = 1.45$. The calculated data are in qualitative agreement with the notions of the dynamics of quantum processes occurring in lasers [8].

Numerical calculations indicate that the pulses of output radiation power reproduce the modulation pulses in shape and duration for modulation frequencies up to 20–25 kHz (Fig. 14.3a). When the transmittance modulation depth is rased above

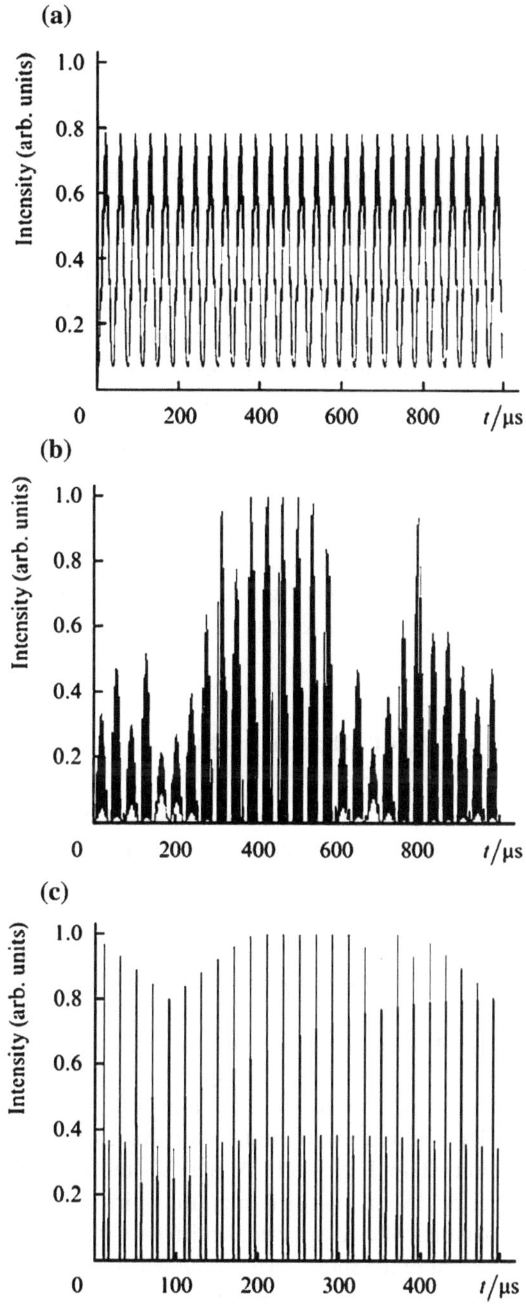

Fig. 14.3 Rough train temporal structures of output laser radiation for a modulation depth of 5 % and a modulation frequency of 27 kHz (**a**), 5.8 % and 27 kHz (**b**), and 5 % and 50 kHz (**c**); $M = 1.45$

the critical value, within the modulation pulse length there emerge separate power peaks, whose total number (4–8) is close to the resonance-to-modulation frequency ratio. Their modulation depth amounts to 100 % (Fig. 14.3b). For modulation frequencies $2/\tau < \nu < \nu_{res}$, the laser goes over to a mode close to the Q-switching mode (Fig. 14.3c). In this case, there occurs not only an increase in the modulation depth of output radiation power, but a change in the characteristic pulse structure—the envelopes of individual pulses and their in tern til peak structure become more regular. The peak intensity in this mode exceeds the stationary intensity by more than a factor of 10. In the CO_2 laser case, the duration of an individual peak of the structure is comparable with the pulse duration of a free- running pulsed CO_2 laser (hundreds of nanoseconds).

Therefore, when the modulation frequency is lower than ν_{res}, the lasing exhibits two characteristics oscillation constituents—the low-frequency oscillation, defined by the modulation frequency, and the high-frequency oscillation, defined by the eigenmodes of the resonator-active medium system, whose frequency is close to ν_{res}. When the modulation frequency is made greater than ν_{res}, the forced oscillations manifest themselves in the form of the high-frequency component, while the slow oscillations are the natural oscillations of the system at the resonance frequency. The results of numerical calculations arc indicative of the feasibility of the repetitively pulsed regime in wide-aperture lasers described by the model (1).

14.4 Experimental Results

The results of numerical calculations were experimentally verified on a test-bench CO_2 gas-dynamic laser whose parameters were given in the foregoing. For a fuel, use was made of carbon monoxide (CO), with air as the oxidiser. The typical output power was equal to 50 kW. To preclude the damage to optical elements of the laser, the output power was lowered by lowering the flow rate of the working components. When the laser was operated in the cw mode, the output power was equal to about 10 kW. Since the dements of the test-bench structure were not cooled, the duration of rims was limited by the heat capacity of resonator elements and combustion chamber and was equal to 3 s, the nominal power settling time was 0.3 s from the onset of mixture combustion.

The optical configuration of the experimental setup was similar to that diagrammed in Fig. 14.1. A part of the output laser radiation (about 20 %) was diverted by an inclined metallic mirror to the injection beam formation system consisting of two spherical mirrors with conjugate focal planes. In the vicinity of the focal plane there formed the waist of the branched part of the laser beam, and a modulator was placed ııar the waist. The modulator location was so selected that the laser beam completely filled the aperture of every round hole in the modulator disk. The modulator was a rotating metal disk with openings dong its perimeter. In experiments, use was made of disks with 150 and 200 drilled holes with respective diameters of 4 and 2 mm and a filling factor (the ratio between the open

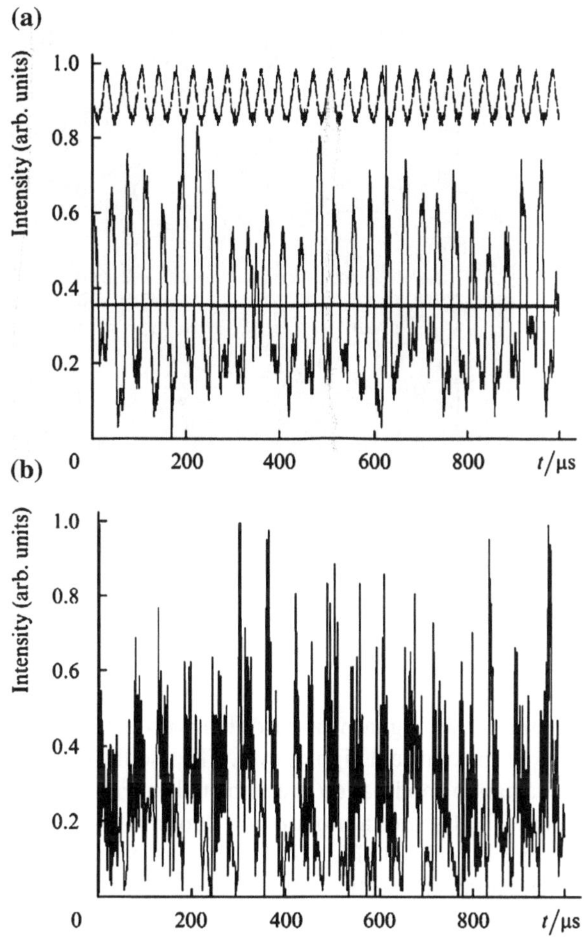

Fig. 14.4 Rough train temporal structures of laser radiation for a modulation depth of ~3 % and a modulation frequency of ~27 kHz (*top* modulating signal, *bottom* output laser signal) (**a**) and for a modulation depth of 7 % and a modulation frequency of ~25 kHz (**b**)

state duration and the total period) of 1:2. The maximal modulation frequency was equal to 33 kHz.

A VIGO SYSTEM PD-10.6-3 photodetector was used as a radiation detector of the measuring system. The detector enabled measuring both the temporal structure of the signal and its constant component. Because the output beam is characterised by a high power density, which is many times higher than the optical breakdown threshold and upper bound of dynamic range of the detector, the radiation was attenuated employing the geometrical factor fa convex deflecting mirror (5) and optical attenuation filters. The signal generated by the photodetector entered a preamplifier. After the preamplifier and a cable line, the signal was recorded by a broadband digital storage Tektronix THS710 oscilloscope The transmission band of the path was limited primarily by the preamplifier and was equal to ~50 MHz.

The average output power was measured with a calorimeter cooled by running water. A mirror 4 focused the radiation onto the calorimeter. In the cw mode, the

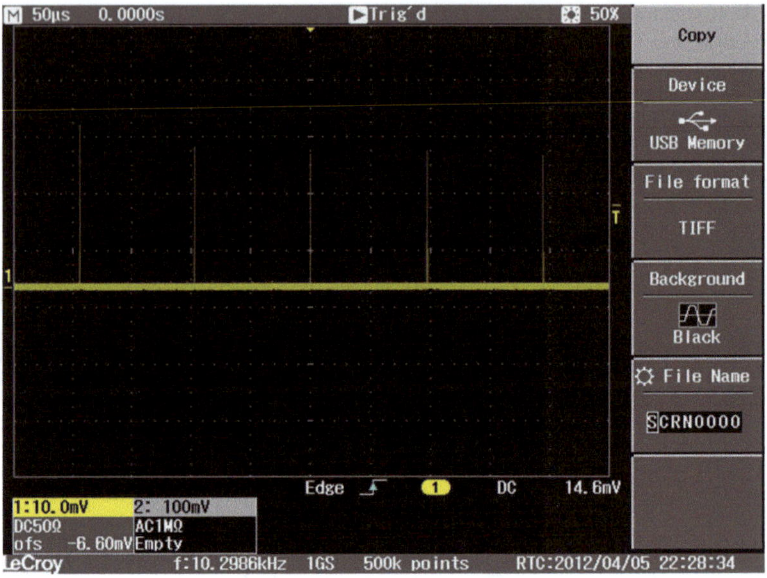

Fig. 14.5 P–P mode of operation for high power Nd YAG laser

constant-level signal was recorded with a noise component, which did not exceed 5 % of the constant signal level. The oscilloscope traces of laser output in the case of radiation modulation are depicted in Fig. 14.4. For a modulation frequency of about 27 kHz and a modulation depth of 2–3 %, the quasi-stationary modulation regime is realised (Fig. 14.4a). In this case, the laser radiation exhibits intensity fluctuations consistent with the modulating signal, with the output power departing from the average value by a factor of three. This regime agrees well with the operating regime shown in Fig. 14.3a. When the modulation depth was increased to 7–8 %, the laser passed to the repetitively pulsed operating regime (Fig. 14.4b). In this case, lasing took place in the form of a train of 5–10 pulses within one cycle of the open modulator state. The duration of an individual pulse was about 200 ns. We emphasise that the recorded pulse duration was limited by the measuring path band-width, which was equal, as noted above, to 50 MHz. The amplitudes of individual pulses exceeded the average value by factors of 6.5–11. This regime agrees well with the regime calculated by expressions (14.1) and presented in Fig. 14.3b. Note that the average output power in the repetitively pulsed regime was equal to the output power in the cw laser operating mode.

The technical characteristics of the modulator in use (the maximal modulation frequency was 33 kHz) did not permit realising the high-frequency modulation modes presented in Fig. 14.3c. Good agreement between the experimental and theoretical data for frequencies ranging up to 25–30 kHz testifies to the adequacy of the proposed model and the possibility of employing this method at higher frequencies to convert a CW laser operation to the operating mode with regular pulses structure. Round trip time of the resonator geometry and P–P regime frequency of pulses matching should be taken into account. The efficient method of regular structure for P–P operating regime for CO_2 and LD pumped Nd YAG lasers has been realized. The studies of the power and temporal characteristics of the laser radiation show that the developed lasers have a very high efficiency of energy extraction close to that of a CW laser mode of operation. The pulse power to average power ratio of more than two orders of magnitude after transformation of CW into P–P mode of operation had been demonstrated. The prospects of efficient and compact P–P/CW laser systems is open (Fig. 14.5).

14.5 Conclusions

We have demonstrated both theoretically and experimentally, the feasibility of converting a high-power wide-aperture CO_2 laser to the repetitively pulsed lasing regime by self-injecting the modulated fraction of output radiation without decreasing the output power compared to the stationary lasing. In this case, the peak output power can exceed the average output power by more than an order of magnitude. Repetitively pulsed modulation with a pulse length of 200 ns to 1 μs, a peak output power greater than 100 kW. and an average output power coinciding with the power of stationary lasing (10 kW) were experimentally obtained. The applicability of the proposed method of laser conversion to the repetitively pulsed lasing regime is limited only by the threshold of optical breakdown on the modulator aperture, which is attained for an average output power of the order of 100 kW in the laser configuration investigated.

References

1. A.M. Prokhorov et al., Proc. SPIE Int. Soc. Opt. Eng. **3574**, 2 (1999)
2. A. Husmann, M. Niessen, F. Grumbel, E.W. Kreutz, R. Poprawe, Proc. SPIE Int. Soc. Opt. Eng. **3343**, 759 (1998)
3. G.D. Hager, B. Anderson et al., Proc. SPIE Int. Soc. Opt. Eng. **4065**, 646 (2000)
4. Y.A. Anan'ev, *Opticheskie rezonatary i lazernye puchki* (Optical Resonators and Laser Beams) (Nauka, Moscow, 1990)
5. Y.S. Vagin, Tr. Fiz. Inst. Akad. Nauk SSSR **113**, 115 (1979)
6. V.V. Apollonov, A.J. Alcock, H.A. Baldis, P.B. Corcum, R.S. Taylor, Opt. Lett. **5**, 333 (1980)
7. V.V. Breev et al., *Energeticheskie diagrammy i issledovanie kharakteristik bystroprotochnykh statsionarnykh CO_2-lazerov* (Energy Diagrams and Investigation of Characteristics of Fast-Flow Stationary CO_2 Lasers) (Izd. IAE im. I.V. Kurchatova, Moscow, 1982)
8. Y.I. Khanin, *Dinamika kvantovykh generatorov* (Dynamics of Quantum Oscillators) (Sov. Radio, Moscow, 1975)

Chapter 15
Mono-module Disk Laser

15.1 Introduction

The mono-module disk laser concept is an effective design for diode-pumped solid-state lasers, which allows the realization of lasers with super-high output power, having very good efficiency and also excellent beam quality. Since the first demonstration of the principle in 1964 the output power of mono-module disk has been increased to the level of few kW in continuous wave (CW) mode of operation. "Zig-Zag" disk laser geometry does not look like as a perspective one for further output parameters growing. The scaling laws for mono-module disk laser design show that the limits for CW mode of operation is far beyond 100 kW for output power and the energy can be higher than 100 J in pulsed mode of operation. Due to the efficient porous cooling technology and possibility of amplified spontaneous emission (ASE) suppression the operation of the mono-module disk laser geometry is possible in CW and pulse-periodical (P-P) modes at extremely high output power.

It has long been customary that as new technologies emerge into the light, potential users and experts start wondering whether these new technologies will replace the old, well-established approaches to the solution of known problems. When we look back at the time when the first laser was created, it becomes clear that each new and more advanced technology usually replaces the pre-existing and well-proven technologies. However, a more differentiated and balanced assessment of the many innovations in the field of laser shows that there is no such thing as a perfect design, or an ideal laser. There is always room for improvement and further development. It is possible that in the near future a laser with the disk geometry of the active medium will become the dominant technology. However, despite this, a number of existing laser technologies (with some exceptions, of course) will continue to improve for quite a long time. Based on these considerations, we must continue to develop most types of lasers, each time clearly specifying their undeniable technological niche and knowing well their advantages and disadvantages.

V.V. Apollonov, *High-Power Optics*, Springer Series in Optical Sciences 192,
DOI 10.1007/978-3-319-10753-0_15

High efficiency and excellent beam quality of disk laser [1] mean that they can be widely used in modern science and industry for a very large range of applications, including treatment of the surface of dielectric materials in microelectronics, cutting, drilling, welding, polishing and cleaning of the surface, other technological operations with superhard and fragile metals and composite materials, etc. Pulse-periodic Q-switched disk lasers and high-average-power, mode-locked laser systems ensure optimal conditions for ablation (sublimation) of a material. A number of advantages of high repetition rate P–P lasers emitting short pulses, as applied to a wide range of industrial technologies, is the basis of many modern concepts of disk lasers. High-intensity light with an insignificant thermal lens effect in the central high-loaded zone of the active medium have led to the lifting of restrictions on the brightness of the pump diode. This has reduced the cost of laser sources, and thus has significantly improved the efficiency of electro-optical conversion, especially in the regime of high average power. The power of the laser source is varied by scaling the cross sectional area of the generated radiation. It should be noted that its ratio to the thickness of the active material in the disk geometry is much larger than the ratio of the cross section of typical core elements of the active medium to their length in any conventional solid-state laser systems. This eliminates the problems associated with nonlinear distortions of the geometry of the active medium, and enables realizing super-high peak values and average powers of laser sources with the disk geometry and the same parameters of the radiation in the far-field region.

15.2 Disk and Fiber Lasers

It is clear that each laser source has its advantages and disadvantages. Beam quality of modern solid-state lasers (disk and fiber lasers) is pretty good, and their efficiency is very high. A fiber laser has significant advantages at low, average and peak output powers, while a disk laser, on the contrary, is preferable in those applications which require high and super-high output powers.

Among high repetition rate P–P sources emitting coherent high average and peak output radiation, disk lasers will undoubtedly dominate today and in the future. They provide the best solution for many industrial applications in the multi-kilowatt power range. A disk is a simple and easily excited laser element, which allows cost-effective generation of radiation with improved parameters. It has a lot of potential for the future. The advantage of the disk laser over the fiber laser is already obvious in the kilowatt power range. Having a large radiating surface, the power density of the disk laser is not critical even at high peak powers. On the contrary, in fiber lasers the increase in the peak power negatively affects the reliability of the resonator. The main disadvantage of the fiber laser is also its high sensitivity to the reflected beam, which often appears in the laser due to the interaction of the generated light with matter. If the radiation reflected from the target affects the fiber laser resonator, then it should be immediately shut down to preserve the cavity.

At the same time, the disk laser resonator is insensitive to the penetration of reflected beams. Even in the case of highly reflective materials, technological operations can be performed without the risk of an emergency switching off of the laser. Another advantage of kilowatt disk lasers is their modular design. This design allows one to replace laser modules during after-sales service, thereby ensuring that in the event of failure of the laser, the downtime as well as the cost of repairs will be minimal.

On the contrary, in fiber lasers, the single-unit design of the resonator makes it impossible to replace easily the failed unit without extensive alteration of the system as a whole. At the same time, the use of fiber laser technology offers certain advantages, especially in the power range (up to several hundreds of watts). This laser is suitable for micromachining of dielectric surfaces and composite materials. However, only the disk laser has a high potential in the case of high peak powers, i.e., when it is necessary to provide a train of short pulses with the highest peak power. In a typical disk laser, as opposed to a fiber laser, the power density inside the cavity is far below the critical damage threshold of the laser active medium and surfaces of auxiliary optical elements, even at a high peak power. For this reason, a disk laser allows, among other things, a better use of the laser medium in general, and significantly improvement in laser efficiency.

Thus, in the future solid-state diode-pumped disk lasers will play the dominant role. Laser diode structures are central elements in many new types of lasers. Today, the decisive factor is the acquisition of the necessary experience and skills to design most efficiently new laser systems. Semiconductor-based technologies and applications of laser equipment on new physical principles are key technologies of modern times. For many years, the world has accumulated the experience needed today in this field of knowledge. The challenge is to ensure technological excellence in the application of laser systems, not only today but in the future.

15.3 Design and Physical Foundations of Disk Laser Operation

The principle of the disk laser operation is based on the use of an active element in the form of a disk with a cooled surface. High cooling efficiency of the laser medium is provided by a large area of the disk surface, which is important from the point of view of the heat transfer process. Therefore, the average power in the beam can reach quite high values. It is important to note that due to effective heat transfer within the disk area there is no thermal lens effect, characteristic of 'rod' and 'slab' (optical range) geometries of the active element. In this case, cooling is performed through a side surface of the active element; a two-dimensional heat flow, forming a parabolic profile of thermal distortion, propagates through this side surface. The temperature dependence of the refractive index in this case leads to the emergence of a strong thermal lens effect. Such a lens deteriorates the directivity of the output and significantly limits the effectiveness of its action. The thin disk, in contrast, is cooled through a thin contact region on one side and generates a one-dimensional axial heat flow on the other side. As a result, the temperature gradient is distributed parallel to the laser beam, which does not result in the thermal lens effect. In practice, a thin laser element in high-power disk lasers is either connected to a porous heat sink or is cooled by forced convection [2, 3].

In both cases, the undesirable thermal lenses are eliminated by the optimum layout of structural elements of the disk laser. Integral heat transfer from the active material to the porous heat sink depends directly on the area of the pump zone and on the geometry of the heat transfer contact. At a constant temperature of the active laser element, the average output power achieved is directly proportional to the area of the pump zone and, consequently, to the cross-sectional area of the laser beam on the disk. This is a unique feature of disk lasers, which allows one to change the power density at a constant energy of the beam on the disk. This property is particularly important for multi-kilowatt sources, because it makes it possible to adjust the value of the peak output power without changing many other parameters that affect the performance of the overall system.

At first glance, the basic challenge is the disk thickness needed for efficient heat dissipation, because the conventional scheme involves low absorbance of the active disk material. However, the pump scheme allows one to find an elegant solution to this problem. The process of excitation in the active material of the disk can be represented in the form of multi-channel optical configuration [4, 5] consisting of a parabolic mirror and a pump deflection system. The parabolic mirror focuses a collimated beam of the laser cavity containing an active element. After partial absorption of radiation passing through the disk body, some energy of the beam is reflected from a highly reflective mirror located on the backside of the disk. As a result of multiple reflections through prisms and the parabolic mirror in general we deal with 20 iterations of radiation passes through the body of the disk, which ensures high efficiency of absorption of light energy of the pump beam. The optical scheme of such resonators tolerates limitations on the brightness of the pump source. Generally, pumping of a high average power disk laser does not require a beam of high optical quality. That is why both laser diodes and rather complex laser diode structures with a uniform intensity distribution can be used in this case. In terms of the practical use laser diodes with a uniform intensity distribution are more preferable due to much lower energy consumption per 1 W of the pump power.

For efficient processing of materials: cutting, drilling, polishing and removal of a thin film from the surface of various materials laser pulses with a high peak power and high repetition rate are needed. Typical pulse duration of a Q-switched Yb:YAG disk laser lies within one microsecond. Q-switching is provided by means of an acousto-optic modulator. Depending on the resonator geometry it is possible to generate pulses with duration from a few hundred picoseconds to a microsecond. To date, the best results achieved with the help of P–P disk lasers having high average power are as follows: pulse repetition rate—up to 10 kHz; maximum pulse energy—100 mJ; peak power—200 kW; average power—up to 1,000 W. However, there are fundamental physical limitations on further scaling of the disk lasers, which we will discuss below.

15.4 Laws of Scaling of Disk Lasers

The power of disk lasers is limited not only by the power of the pump and overheating of the medium but also by the losses due to the ASE and the background radiation losses in the resonator. To avoid overheating, the size of the active medium should vary in accordance with the law of power scaling. Then, to avoid large losses caused by their exponential growth during the ASE, amplification of radiation corresponding to transverse round trips should not be large. This requires a reduction in gain G. Gain is determined by the reflectivity of the output mirror and disk thickness. Amplification of radiation per round trip, however, should not be substantially greater in magnitude than the radiation loss per round trip along the same optical path. The difference between the gain and losses per round trip of radiation determines the optical energy that is coupled out from the laser cavity. Reduction of the gain at this loss level requires an increase in the disk thickness. In this case, at a certain critical size, the disk becomes optically too thick and cannot be pumped above the threshold without overheating. Some features of scaling can be shown on a simple model. Suppose M is the saturation intensity of the medium. The corresponding optimal thickness of the disk can be estimated as h~T/Mb. The corresponding optimal lateral dimensions of the disk can be represented in the form: D~T/Mb2, where T is a parameter of the heat load. Roughly speaking, the loss per round trip of radiation must be scaled inversely proportional to the cube root of the power required: P~nT2/Mb3.

An additional problem is the efficient delivery of pump energy. In cases with a small-signal gain, single-pass absorption of the pump is also low. It follows that the effective utilization of the pump energy is very necessary for the effective operation of the laser disk. To scale the output power, the medium must be optically thin, which requires a large number of passes of the pump energy through the medium. Besides, the pump energy coupled in through the lateral side of the disk can also be a possible solution for efficient pumping.

To reduce the effects of the ASE, it was proposed to use an optical cover consisting of an undoped material on the surface of the laser disk. This cover allows the spontaneously emitted photons to escape from the active layer and prevents their resonance in the bulk of active material. Rays cannot be reflected from the surface, as in the case of an open disk. This allows the maximum power achievable by the disk laser to be increased by an order of magnitude. Reflection of the ASE from the disk edge should also be suppressed. This can be done through the absorbing layer at the generatrix of the disk cylinder. In the regime when output power is close to maximal, much of the energy is used in the ASE; therefore, the absorbing layers must also have radiators accumulating heat. In the case of the maximum pump density of the disk laser its efficiency is quite low: Most of the pump power is used in the ASE and is absorbed at the edges of the device. In this case, the distribution of the pump energy between several disks can significantly improve the performance of the laser system. Indeed, lasers, consisting of several modules with disk elements in a single cavity, have been repeatedly reported. One

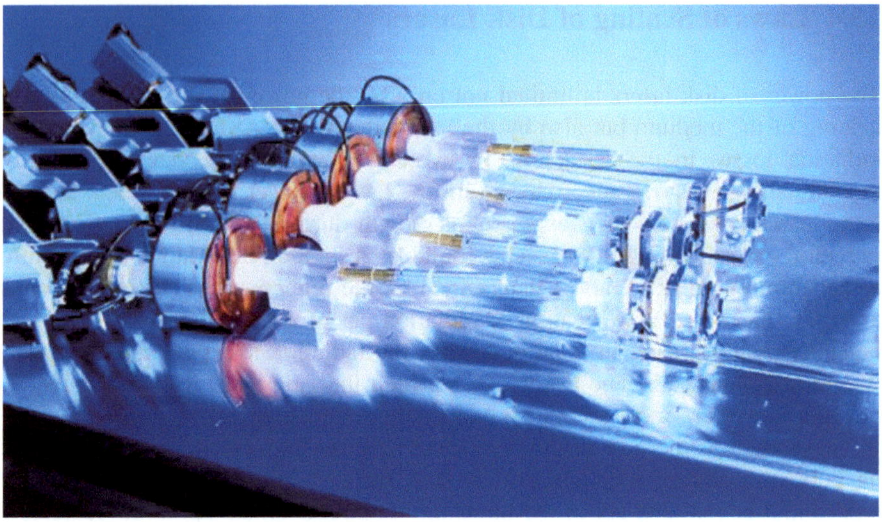

Fig. 15.1 "Zig-Zag" disk Laser consisting of a series of disk modules in a single cavity, TRUMPF [5]

of these lasers, fabricated by the TRUMPF Group [5]—a world leader in this class of laser systems—is presented at Fig. 15.1.

In the history of electrical discharge CO_2-lasers output power scaling a very similar geometry had been investigated carefully. This is a "Zig-Zag"—CO_2-laser which was created in FIAN by A.I. Barchukov in the beginning of 1970s. The future of "Zig-Zag" multi-module geometry disk laser probably will be the same. At the Fig. 15.2 "Zig-Zag"—CO_2-laser created by A.I. Barchukov FIAN 1970 [2] is presented. With the help of this laser we have forecasted a lot of new effects which later on made significant difficulties to the scientists involved into high power mono-module laser creation. There are: deformation of the high power laser

Fig. 15.2 "Zig-Zag"—CO_2-laser created by A.I. Barchukov FIAN 1970 [2]

mirrors surface and distortion of phase front of the beam during propagation and as the result—high-power static and adaptive optics, big scale water cooled optics for telescopes based on SiC, high power CW plasmotrone in air and as the result—crusibleless growing of crystals, high quality laser beam and as the result—high quality laser based hard materials welding and cutting and so on.

In the quasi-continuous regime, the power can be estimated by scaling the saturation intensity with a duty cycle of the pump and by multiplying the duration of the pump by the pulse repetition rate. At moderate repetition rates (e.g., higher than 1 Hz), the maximum energy of the output pulses is approximately inversely proportional to the cube of background losses b. The undoped cover can increase the average output power by an additional order of magnitude, provided that this cover does not increase the background loss. At low pulse repetition rates (in the single-pulse regime) and sufficient pump power, there is no general limitation on the energy, but the required size of the device increases rapidly with increasing pulse energy, thereby establishing a practical limit of energy. One active element, according to the estimates, can generate an optical pulse with energy of a few thousand joules, depending on the level of internal signal loss in the disk.

Disk lasers, as well as fiber lasers, have a large ratio of the cooling surface area to the gain of the laser. However, these two different concepts of the laser design also differ by the values of the achievable peak power. Beam quality of fiber lasers is determined by properties of waveguides, the refractive index difference between the core and the cladding, and the size of the internal diameter of the fiber that transmits light. On the other hand, beam quality of disk lasers depends on the design of the resonator. With increasing size of the optical pump region (its diameter is usually a few millimeters) at a constant optical power density on the disk surface, scaling of the output power becomes possible. Adjustment of the resonator also has similar features. Self-phase modulation defines the typical limit of nonlinear amplification of ultrashort pulses. It leads to an extension of the spectral line proportional to the ratio of the effective optical path in the material with nonlinear characteristics to the effective cross-sectional area of the beam multiplied by the square of the pulse duration. An electro-optical switch in a regenerative amplifier of the disk laser makes a substantial contribution to the nonlinear characteristics of the gain which is greater than the contribution from the disk. It must be emphasized that the values of the peak power, achieved today in single disk modules, are much smaller than the values obtained in 'rod' and 'slab' solid-state laser systems. However, the level of the average power generated by a single disk module also leaves much to be desired.

15.5 Regenerative Amplification of Pulses

At present, in the scientific and technical research and manufacturing processes relying on the use of high average power lasers, only sources operating in two regimes find applications, i.e., CW and P–P laser sources with a pulse repetition rate up to hundreds of hertz and pulse duration from tens of microseconds to tens of milliseconds. In most processes, purely thermal mechanism of action is mainly implemented, since use is made of the possibility of a laser source to deliver quite a lot of energy to a small area of the surface of the work piece. High-frequency, high-average-power laser systems operating in the Q-switched regime, which provides the pulse duration in a periodic train from a few to hundreds of nanoseconds, allows for a fundamentally different mechanism of interaction of radiation with matter, i.e., ablation, ensuring local release of energy not only in space but also in time [6]. This results in an explosive local evaporation of the substance without an intermediate liquid phase. This mechanism can significantly extend the range of technological applications of laser sources. Today, it has found a real use in practice, only at an insignificant level of output power in the range of 1,000 W. However, there arises a reasonable question: Why are high-power P–P lasers not available at the international laser market? The answer is simple: The complexity of their manufacturing consists in the inability of application of the classical modulation methods to laser systems with high average and peak powers. At the same time, it is quite clear that the creation of high-power (>1 kW) P–P lasers with frequencies up to 100 kHz or higher (with a peak power exceeding the average power by two to three orders of magnitude) would significantly expand the field of application of laser sources, increase their efficiency, and implement qualitatively new effects [7].

One of the successful trends in implementation of new methods of high-frequency modulation of the output from high-power laser sources is due to the regenerative amplification of a small signal injected into the cavity of a high-power gas or solid-state laser with the classical rod geometry. Use of intracavity modulators in high-power laser systems is hindered, because high power density inside the cavity leads to the appearance of plasma on the surface elements of the modulator, to the screening of radiation and to the destruction of the optical elements of the modulator. Quite promising is the method of realization of the P–P regime in high-power lasers, which is based on the use a self-filtering cavity. This cavity is a confocal resonator consisting of two spherical mirrors of different curvatures; in the common focal plane there is an annular output mirror with a hole. Because of the high degree of discrimination of the higher modes with respect to the resonator losses, the lower mode is reliably generated. The spherical resonator mirror is placed outside the active medium and contains a modulation unit comprising spherical mirrors. The laser mode should be formed quickly enough in the leading edge of a giant pulse. The problem of laser micromachining usually involves cutting, drilling, surface cleaning and polishing, as well as removal of thin layers due to the ablation of the material. Micromachining relies on lasers emitting pulses of different duration. The geometry of the beam and the pulse repetition rate are crucial: usually the pulse duration does not exceed a few tens of nanoseconds, and in some cases reaches several hundreds of picoseconds. Disk laser technology provides excellent flexibility in the scaling of the output power by combining it with a high pulse energy and excellent beam quality.

15.6 Prospects for Scaling the Power of Disk Lasers

As mentioned above, the design of the disk laser generating a continuous and relatively high power is ideal for cutting and welding metals, where high optical quality of the beam is required. It is important for such industries as automotive, transportation, aerospace and heavy engineering. However, the design of the disk laser is more suitable for a range of new technologies that are currently in demand. Today, a multi-module laser with a power up to 16 kW and beam quality of ≥2 mm mrad, fabricated by the TRUMPF Group, is the undisputed leader in this class of laser systems. CW solid-state diode-pumped disk lasers demonstrate highly efficiency. Beam quality of the laser disk is outstanding, which makes it possible to work with a target from a large distance, while providing an extremely high concentration of radiation in the interaction region with the help of the focusing optics. In this design of the laser module the disk size is no more than 4−5 mm, because at larger sizes the energy loss due to the amplified spontaneous emission increases at an absolutely unacceptable rate. The above-presented laser system consists of a set of disk modules in a single cavity. This geometry of the laser system allows implementation of the P–P regime with a fairly high peak power, but the average power lies within a kilowatt. Parallel operation of disk lasers can increase the peak power of the entire system, but the phase-locking of the disk channels in the P–P regime requires additional scientific research. Further expansion of the power to the megawatt level of both the average and peak power in both geometries is very problematic.

At the same time, a different approach is known to the implementation of scalable solid-state laser systems, which consists of a set of active elements in the form of 'slabs', followed by phase-locking of the generated radiation. The team of the Northrop Grumman Corporation has created a laser with a power of >100 kW and high laser beam quality equal to the diffraction limit of 1.5 (averaged value) with the exposure time of 300 s [8]. The laser efficiency reaches 30 %. The authors of the project point to the easy rcplacement of individual laser channels in case of failure. They also mentions some of the advantages of the parallel structure of the amplifying channels in terms of ease of further increasing output power, if necessary, to 100 kW. Beside, the laser assembled according to this scheme provides the divergence at the level of two diffraction limits from the common (composite) aperture. From general physical considerations, we can assume that for a given power level and a reasonable value of radiation resistance of optical elements, the divergence of the laser radiation at the level of 2×10^{-5} rad can be achieved for the CW generation regime and 0.6×10^{-5} rad for the P–P regime. A further increase in the number of channels in order to obtain an output power of 1 MW will require the coherent summation of the power from at least 80 channels, which seems to be an elusive task.

The question arises as to how the average power of a few megawatts can be achieved on the basis of solid-state lasers. And it is this power that is needed to address many problems associated with the removal of debris from near-Earth space, with launching of missiles with the help of lasers, with the creation of long-range conducting channels and others. Fiber lasers are not applicable for these purposes because of the smallness of the area of the exit pupil of the fibers and hence the impossibility of operation of such lasers in the high-frequency P–P regime with high peak power at an average power of a few megawatts [9, 10]. The laser system based on 'slabs' also seems hard to implement as adjustment of the system and its maintenance in a safe operation mode are comparable to the complexity of working with a multi-element system for the solution to the problem of controlled thermonuclear fusion (CTF) at Livermore (USA). Proceeding from the above, the answer is quite clear: Such a laser can and should be based on the mono-module disk laser geometry!

This is due to the indisputable advantages of the disk geometry in terms of the minimal thermal lens in the active media and the high radiation resistance of the disk in the P–P regime because of the large area of the optical surface to couple out the radiation. So, the necessity to find a solution to the problem of the ASE suppression along the diameter of the disk was the major problem (matter of patent). In our case, the size of the disk at a megawatt level of the average power output should be at least 50 cm, i.e., at least hundred times bigger the size of the disk that is used today in existing systems. Radiation from such a laser, obtained during generation in the active medium of a single disk, does not require additional phase-locking. At the same time, such a laser in a mono-module geometry will be very well combined with a large-diameter telescope for ensuring high peak power density of the laser pulse on space debris. It is well known that the disk geometry of a laser was proposed 51 years ago [11], but only now the solution to the problem of the ASE suppression with increasing transverse dimensions of the active medium in the mono-module disk geometry is found! Thus, the prospects of new versions of the mono-module disk laser creation for new class of cutting edge problems of our days are open [12–15]!

15.7 Conclusions

Only the future will show the final physical-technical solution for the mono-module disk laser which will be used effectively for the processing of materials and for the solution of other important challenges faced by science and technological advances of the future. However, today we can say that creation of megawatt class high repetition rate P–P disk lasers with a large cross section of the active medium will open up a great prospects for their use in solving the problems of launching of small satellites with lasers, formation of super-long conducting channels in space and atmosphere, cleaning of the near-Earth space from the space debris, etc.

References

1. N.G. Basov, O.V. Bogdankevich, A.Z. Grasiuk, IEEE J. QE **2**, 9 (1966)
2. V.V. Apollonov, A.I. Barchukov, A.M. Prokhorov, JEEE J. QE **6**, 10 (1974), Report of FIAN (1971)
3. V.V. Apollonov, A.M. Prokhorov, A.H. Guenther, Laser Phys. V **11**, N.8 (2001)
4. A. Giesen, H. Hugel, A. Voss, K. Wittig, U. Brauch, H. Opower, Appl. Phys. **B58**, 365 (1994)
5. SPIE Events Europe, TRUMPF's R@D (2012)
6. V.V. Apollonov, P.B. Corkum, R.S. Teylor, Appl. Phys. Lett. **35**, 147 (1979)
7. V.V. Apollonov, V.I. Kislov, V.V. Kijko, A.G. Suzdal`tsev, QE **33**, 9 (2003)
8. SPIE Events Europe, Nothrop's R@D (2012)
9. V.V. Apollonov, Symposium HPLS@A-2012, Istanbul, September 10 (2012)
10. V.V. Apollonov, V.I. Kislov, A.M. Prokhorov, QE **23**, 5 (1996)
11. A.Z. Grasiuk, QE **42**, 12 (2012)
12. V.V. Apollonov, Nat. Sci. **4**, 713 (2012)
13. V.V. Apollonov, Nat. Sci. **4**, 719 (2012)
14. V.V. Apollonov, Nat. Sci. **4**, 725 (2012)
15. V.V. Apollonov, Am. J. Mod. Phys. **1**, 1 (2012)

Chapter 16
Pulse-Periodic Lasers for Space Debris Elimination

16.1 High Repetition Rate P–P Mode of Laser Operation

For the lasers with a high average output power (GDL, HF/DF, COIL, Nd YAG) is very common to use an unstable resonator configurations having a large cross section of the active medium. In the resonators of this type, externally injected low-power beam may exert a significant effect on the characteristics of output radiation.

One way to realize the radiation control regime is the self-injection regime of radiation, extracted from the resonator and returned back to resonator as a part of radiation after changing its spatial-temporal characteristics [1, 2]. The transition to the transient lasing mode is effected through the modulation of the self-injecting beam. Earlier, a study was made of laser versions with radiation self-injection into the paraxial resonator region. However, analysis showed that the power of the beam injected into the paraxial beam region should be about the same value or comparable with the output laser power to efficiently control the resonator of a continuously pumped laser, unlike pure pulsed systems with regenerative amplification.

The self-injection of a part of output radiation through the resonator periphery is more efficient: on return to the paraxial resonator region, the injection power significantly rises due to the large number of passages to play the dominant part in the formation of output radiation.

In the case of a traditional resonator, the role of waves converging to the resonator axis was found to be insignificant, because their source is a narrow region with a small relative area at edge of the output mirror; accordingly, the power of the control wave injected into the resonator is low. This wave has a large divergence, and only its small part (of the order of 1/Nf, where Nf $\gg$ 1 is the Fresnel number) participates in lasing.

The effect of injection wave on the resonator characteristics can be enhanced by matching the beam phase with the resonator configuration and increasing the radiation power returned. In this case, the propagation direction and the wave front

V.V. Apollonov, *High-Power Optics*, Springer Series in Optical Sciences 192,
DOI 10.1007/978-3-319-10753-0_16

curvature of the injection beam should be so matched with the resonator configuration that the injection beam concentrates, after a relatively large number of passages through the resonator, near the optical resonator axis and transforms to a divergent wave that forms the output radiation. The injection beam energy should be high enough to exceed, after its arrival to the resonator axis, the saturation energy of the active medium.

P–P mode of operation was realized in two type of lasers theoretically and experimentally, in a gas-dynamic CO_2 and Nd YAG lasers [3]. CO_2—laser had the following parameters: the length of the active medium $L_a = 1.2$ m, the unsaturated gain coefficient $g_0 = 0.6$ m^{-1}, the time it takes the active medium to transit the resonator $\tau = 0.92 \times 10^{-4}$ s, the relaxation time $\tau_r = 2.76 \times 10^{-4}$ s , the total go-round resonator time $\tau_{rt} = 4.2 \times 10^{-9}$ s, the luminescence lifetime $\tau_{\text{lum}} = 5$ s, the resonator magnification factor $M = 1.45$, the diameter of output laser aperture $a = 0.08$ m. Nd YAG—laser was above of 1 kW level, with two heads geometry.

The CO_2-laser resonator is made up of two spherical mirrors with rectangular apertures, which provided a geometrical amplification factor of 1.45. The active medium travels across the optical resonator axis. In what follows all theoretical and experimental data are given for a laser with the above parameters.

A part of the output laser radiation was diverted by an inclined metallic mirror to the injection beam formation system consisting of two spherical mirrors with conjugate focal planes. In the vicinity of the focal plane there formed the waist of the branched part of the laser beam, and a modulator was placed near the waist. The modulator location was so selected that the laser beam completely filled the aperture of the modulator. The maximal modulation frequency in our experiments has reached 50 kHz.

A mirror 4 focused the radiation onto the calorimeter. The duration of an individual pulse was about 100–150 ns. We emphasize that the recorded pulse duration was limited by the measuring path bandwidth equal, as noted above, to 50 MHz. The amplitudes of individual pulses exceeded the average value of output power by factor -10. The average output power was measured with a calorimeter cooled by running water. It is noteworthy that the average output power in the pulse-periodic mode was equal to the output power in the CW laser-operating mode. Good agreement between the experimental and theoretical data for frequencies ranging up to 25–50 kHz testifies to the adequacy of the proposed model and the possibility of employing this method at higher frequencies to convert a CW laser to the operating mode similar to the Q-switching mode.

HF/DF-laser and COIL are waiting for the experimental efforts to be applied. Theoretically P–P modes of regenerative amplification for high power lasers have been investigated and modeled by computer. The output parameters are very much dependable on parameters of media, way of pumping and resonator geometry. The summary of the radiation temporal structure is presented:

COIL	P–P mode starts at frequencies >20 kHz
	Depth of modulation—100 % at frequencies >100 kHz
	Pulse duration <250 ns. Ratio P peak./P aver. = 100–1,000
HF/DF	P–P mode starts at frequencies >100 kHz
	Depth of modulation—100 % at frequencies >250 kHz
	Pulse duration <100 ns. Ratio P peak./P aver. = 100–10,000
Nd YAG	P–P mode starts at frequencies >4 kHz
	Depth of modulation—100 % at frequencies >40 kHz
	Pulse duration <250 ns. Ratio P peak./P aver. = 100–1,000
CO	P–P mode starts at frequencies >10 kHz
	Depth of modulation—100 % at frequencies >100 kHz
	Pulse duration <250 ns. Ratio P peak./P aver. = 100–1,000

16.2 New Application for Space

Last years the increasing of attention has been given to the study of possibility of lasers use for cleaning of the space from elements of space debris (ESD) which have collected within more than four decades of operation of space and create in some cases the big threat for space vehicles (SV). By expert estimations in the space by 1996 was about 3.5 million not traced ESD in the size less than 1 cm, more than 100,000 splinters in the size in a diameter from 1 to 10 cm, the size nearby 8,000 ESD exceed 10 cm [4–6]. Large ESD with a diameter more than 10 cm are found out by modern watch facilities and are brought in special catalogues. The most effective method of protection from such ESD is maneuvering of SV. By estimations of experts splinters in diameter less than 1 cm do not represent special danger for existing SV because of presence of passive constructional protection though it considerably makes SV much more heavier. The most unpleasant diameter of splinters is 1 … 10 cm when the necessary degree of passive protection does not manage to be carried out because of its unacceptably big weight, and to avoid collision at the expense of maneuvering SV it is impossible, as on the radar screen such splinters are not visible.

In low orbits under the influence of atmosphere quickly enough there is their self-clearing as time of life ESD in orbits with height about 200 km averages about 1 week. In higher orbits in height their self-cleaning can occupy of 600 km 25–30 years, and at heights about 1,000 km—2,000 years [5]. The estimations executed in work [6], have shown that the probability of collision SV in diameter of 10 m within 1 year of its operation makes 0.45×10^{-2} for ESD with a size 2 … 4 cm and 0.4 for ESD with a size <0.4 cm, and frequency of collisions with the catalogued objects ($\geq$10 cm) is at level of one collision for 30 years. And every year number ESD regularly increases. From here the reality of threat of collision with ESD for all period of operation SV is clear.

Thus withdrawal ESD from an orbit protected SV is rather actual problem. For this purpose it is necessary to reduce speed of its movement. As it will be shown further, it is possible to reach at the expense of pulse irradiation ESD and reception on its surface of the plasma creating an impulse of return. Such impulse arising in a mode of laser ablation of ESD material, should reduce height of orbit ESD so that it has flown by SV or, finally, would enter into dense layers of atmosphere and has burnt down.

In a number of works earlier it was offered to use the Nd YAG ground based laser installations for space clearing, but such laser has the essential lacks connected with necessity of passage of 1 mkm radiation of the big capacity through atmosphere that can lead to loss of optical quality of a bunch of radiation and occurrence of nonlinear effects. They have small mobility, therefore number ESD which it is possible to influence on by their radiation, will be limited. At influence of laser on ESD from the Earth surface the return impulse will be directed upwards and the apogee of orbit ESD will be increased, but the perigee is going to be decreased and stopped by dense atmosphere. And the most important thing—requirements to power of the land based laser should be increased in comparison with the space laser as the distance from a terrestrial surface to ESD is much more. For these reasons for the most expedient arrangement of Nd YAG 100 kW laser installation directly in space is recommended. Thus it is desirable, that in such installation power consumption should be minimal. To this condition satisfies Nd YAG with LD pumping, capable to work independently in P–P mode of operation with very small expenses of power for the system service control. But for the case of the ground based laser we have suggested the DF-laser, which radiation propagation through the atmosphere is much more effective and the output power of existing systems (>1.5 MW) and technology are more advanced.

In the paper [5] the most possible variants of rapprochement ESD, flying, as a rule, on elliptic orbits, with various SV, moving on circular orbits at heights 200–700 km have been analyzed. Two variants when SV moves on a circular orbit at height of 400 km have appeared the worst, and ESD fly on elliptic orbits with height of apogee −2,000 and −4,000 km. In this case in a perigee there are areas where planes of orbits SV and ESD coincide, and speed of their rapprochement is maximum, and in this area vectors of speed SV and ESD lie along the same direction, i.e., by influence of the laser radiation it is impossible to give to ESD a lateral component of speed, as in more opportunity of an inclination of planes of orbits SV and ESD under the relation to each other.

The maximum speeds of rapprochement calculated for these two variants have made accordingly −395 and −2,463 m/s. For circular orbits with height 200, 400 and 700 km settlement of rapprochement ESD speeds, flying on circular orbits, with SV do not exceed 343 m/s, therefore these variants can be neglected.

Let's consider process of rapprochement ESD which is catching up SV, after influence on ESD of the laser radiation. Before the laser influence force of an attraction of the Earth and centrifugal force are equal:

$$\frac{mv_0^2}{R+H} = \gamma \frac{mM}{(R+H)^2},$$

where υ_0—speed of movement ESD on a trajectory before influence of a laser impulse, R—radius of the Earth, H—height ESD over the Earth, M—weight of the Earth, γ—a gravitational constant, m—weight ESD after such influence on ESD this balance will be broken; then the reduction of ESD speed $\Delta\upsilon$ will force the normal acceleration in the direction to the centre of the Earth:

$$a_{\mathrm{H}} = -\frac{\gamma M}{(R+H)^2} + \frac{(\upsilon_0 - \Delta\upsilon)^2}{R+H}$$

where $\Delta\upsilon$—change of ESD speed after laser pulse influence (typical value $\Delta\upsilon$ makes ~200 km/s [7]). After simple transformations from (1) we will receive

$$a_{\mathrm{H}} = \frac{-2\cdot\upsilon_0\cdot\Delta\upsilon + \Delta\upsilon^2}{R+H}.$$

Through the time—t the radius-vector of ESD orbit will be changed:

$$\Delta H = \frac{a_{\mathrm{H}}t^2}{2} = \frac{\Delta\upsilon\cdot(\Delta\upsilon - 2\upsilon_0)}{R+H}t^2$$

By knowing the initial distance—L from SV to ESD and a tangential component of rapprochements speed of ESD to SV after influence of a laser pulse:

$$t = \frac{\ell}{\upsilon_{\mathrm{T}}} = \frac{\ell}{\upsilon - \Delta\upsilon}$$

Then for change of size of a radius-vector of ESD orbit we will receive the following expression:

$$\Delta H = \frac{\Delta\upsilon\cdot(\Delta\upsilon - 2\Delta\upsilon_0)}{R+H}\frac{\ell^2}{(\upsilon - \Delta\upsilon)}$$

From here it is possible to find the distance between ESD and SV when it is necessary to start the influence on ESD by laser:

$$\ell = (\upsilon - \Delta\upsilon)\sqrt{\frac{\Delta H\cdot(R+H)}{\Delta\upsilon\cdot(\Delta\upsilon - 2\upsilon_0)}}$$

Proceeding of the SV dimensions, we will be set by size $\Delta H = 30$ m. Then for the first variant of rapprochement ESD to SV at V rapp = 395 m/s, $\Delta\upsilon = 200$ m/s, Vo = 8 km/s, $H = 400$ km, R ≈ 6,300 km the distance between will make 4.1 km. This way will be passed in time ~20 s. Then for metal ESD at typical values $C_{\mathrm{m}}^{0\Pi\mathrm{T}} = 4\,\mathrm{din} - \mathrm{s/J}$ and $S/m = 0.15$ sm2/g we will receive $\Delta\upsilon = 6\mathrm{cm/s}$, and the necessary number of pulses for the value $\Delta\upsilon = 200\,\mathrm{m/s}$ will make 3,300 pulses at frequency of high repetition rate Nd YAG laser—3,000 Hz necessary time of influence 1.1 s, which is much less than time of rapprochement to SV found before (20 s). It shows that with the same laser it is possible to reject ESD from SV with rapprochements having much greater speed.

For more exact calculations at the big speeds of rapprochement it is necessary to consider dynamics of change of values $\Delta \upsilon$ and current distance between ESD and SV after influence of each laser pulse of P–P irradiation of ESD.

For the second variant with very great speed of rapprochement V rapp = 2,463 m/s at $\Delta \upsilon = 200$ m/s and much bigger distance—20 km the rejection is possible as well. However, maintenance of $\Delta \upsilon$ at the distance—20 km will meet some changes of parameters due to the bigger size of the focal point on such a distance.

The problem of ESD withdrawal from SV orbit that ESD has flown by SV has been above considered. Other problem is important also—to create such impulse of return to achieve decrease ESD to an orbit in height of 200 km when at the expense of the further braking in atmosphere of particles of ESD will be burn down, and the space will be cleared from ESD. In other words, SV with laser installation will carry out onboard a role of "cleaner" of the most used orbits. That particle ESD has decreased to 200 km over the Earth surface, its speed needs to be reduced by certain value $\Delta \upsilon$ which will allow it to pass from a circular orbit on elliptic which exact value can be calculated as follows:

$$\Delta \upsilon = \mathrm{V}_{\mathrm{apogee}} - \mathrm{V}_{\mathrm{start}},$$

where Vapogee—ESD speed in apogee of a transitive elliptic orbit, V start—speed ESD in an initial circular orbit. Speed in apogee is:

$$\mathrm{V}_{\mathrm{apogee}} = \sqrt{\frac{2 \cdot \gamma \cdot M \cdot R_{start}}{r_{200} \cdot (r_{200} + R_{start})}},$$

where R 200—radius of a circular orbit in height of 200 km, R start—radius of an initial orbit. ESD speed in an initial circular orbit is defined as:

$$\mathrm{V}_{start} = \sqrt{\frac{\gamma \cdot M}{R_{start}}}.$$

On the basis of given by [5] data, the graphic dependence of demanded reduction of speed ESD in apogee of an elliptic orbit from height of an initial circular orbit has been constructed. The similar dependence has been resulted in the work [5] without explanations. It is clear that ESD, being in the orbit with height ~900 km, will decrease it to the height of 200 km if to reduce the speed by 200 m/s.

Change of ESD speed $\Delta \upsilon$ after the influence of laser radiation pulse with energy density E [J/cm^2] on ESD is defined from the following expression:

$$\Delta \upsilon = \mathrm{Cm\,E\,S/m};$$

where S—the interaction area, m—weight of ESD, Cm [din s/J]—proportionality factor between $\Delta \upsilon$ and E, depending on the ESD type. Characteristics of the most widespread of them are presented in the Table 16.1 [5]. Such ESD are formed as a result of SV explosions, or their collisions with ESD. Spheroids of Na and K are formed after destruction of reactors, splinters of phenol-carbon plastics and

Table 16.1 Cm(opt) and S/m for different ESD

	Type Of ESD				
	Na(K)	"C"-based materials	Organics-based materials	"Al"-based materials	"Fe"-based material
Angle (degre.)	65	87	99	30	82
Apogee (km)	930	1,190	1,020	800	1,500
Perigee (km)	870	610	725	520	820
S/m (cm^2/g)	1.75	0.7	2.5	0.37	0.15
size (cm)	1.0	1×5	0.05×30	1×5	1×10
Reflectivity	0.4	0.02	0.05/0.7	0.05/0.7	0.5
$C_m^{opt}\left(\frac{\text{Din s}}{\text{J}}\right)$	(6 ± 2)	(7.5 ± 2)	(5.5 ± 2)	(4 ± 1.5)	(4 ± 1.5)

Table 16.2 LEO /MEO ESD removal data for Nd YAG laser

λ	τ	D_b	W	f	<P>	d_s	Z	I
1.06 μm	10 ns	30 m	60 J	3,000 Hz	360 kW (0.5)	5.2 cm 2 Dif	300 km	3.0 J/cm^2

fragments of "plastics-aluminum" are the fragments of thermal protection; splinters of aluminum based materials can appear after explosion of tanks and covers of SV; steel bolts—fragments of connecting blocks armature.

High power high repetition rate P–P laser should generate a temporally, and spectrally effective pulse designed for high transmission through the atmosphere, as well as for efficient ablative coupling with the target.

The space based Nd YAG laser with output power less than 100 kW that we propose is the best tool for fast re-entering of the ESD to the dense layers of atmosphere.

The DF ground-based laser system that we have proposed is capable to get a rapid engagement of targets whose orbits cross over the site, with potential for kill on a single pass. Very little target mass is ablated per pulse so the potential to create additional hazardous orbiting debris is minimal.

The laser system would need to be coupled with a target pointing and tracking telescope with guide-star-like wave-front correction capability.

Table 16.2 presents the LEO /MEO ESD removal data for Nd YAG laser. ESD have a size 1–10 cm and fly below 300 km altitude. Cm = 4 dyn-s/J in average for polymer and "Al"—based materials response. Typical S/m data for ESD: NaK-1.75; Al-0.37; Fe-0.15 are taken from the Table 16.1. For I = 3.0 J/cm^2, S /g = 0.15 cm^2/g, we need N = 7,000 laser pulses for ESD re-entry. Nd YAG—laser operating at 3,000 Hz can re-enter small object from the gap 1–10 cm in 2.3 s. Such a level of average output power (360 kW) for CW/P–P Nd YAG lasers has not been demonstrated up to now. To get such effective results for clearing we need not the laser only but a 30 m in diameter telescope to deliver the laser pulses to a target at 300 km range or more with 10 ns time duration:

Table 16.3 presents the LEO /MEO ESD removal data for DF laser. ESD have the same size 1–10 cm and fly below 300 km altitude. Cm = 4 dyn-s/J in average

Table 16.3 LEO /MEO ESD removal data for DF laser

λ	τ	D_b	W	f	<P>	d_s	Z	1
3.8 μm	10 ns	30 m	0.150 J	10 kHz	1.5 MW	18 cm 2Dif	300 km	0.6 J/cm^2

for the same materials: polymer and aluminum. With I = 0.6 J/cm^2, t = 10 ns, S/g = 0.15 cm^2/g, we need N = 35,000 pulses for ESD re-entry. Ground based 1.5 MW DF—laser operating at 10 kHz can re-enter any small object from the same size gap in 3.5 s. This operation requires a 30 m in diameter telescope to deliver 2 J/cm^2 (Cm = 0.2 Cm opt) to a target at 300 km range with a 10 ns pulse at 3.8 mkm: Here is important to note that with 1 min delay for retargeting all objects of this height and below can be re-entered during—0.5 year only. It should be noted as well here that the level of output power for CW regime had been demonstrated and technology is mature enough. The realization of P–P mode of operation for this type of laser is the question of time. Motivation is completely available. New tasks for high repetition rate high power lasers generated during last few years are very much important [7–9] and definitely should be solved during upcoming years.

16.3 Conclusion

1. In the paper the questions of SV protection and of orbits clearing from dangerous ESD with diameter from 1 to 10 cm by means of high power high repetition rate P–P Nd YAG with average power 100 kW and DF-lasers with average power about 1.5 MW are considered;
2. Possibility of applying mentioned above installations not only for dangerous ESD withdrawal from SV orbit, but also for planned clearing of the most maintained orbits from these ESD when the given installations will carry out a role of "cleaner" of these orbits is analyzed. For this purpose under the influence of radiation it is necessary to translate ESD from a circular orbit to elliptic, which perigee is in the dense atmosphere beds where ESD should be burn down. As a result of the decision of a ballistic problem, dependence of necessary reduction of speed ESD from height of their orbit over the Earth is received. It is found that for orbits with heights up to ~300 km the demanded influence can be provided by 1.5 MW DF-laser installation with a telescope with diameter −30 m and duration of pulses about 10 ns;
3. It is shown that for the worst variant in case of influence on metal ESD with the greatest speed of their rapprochement with SV ~2.5 km/s angular divergence of radiation of space based 100 kW Nd YAG laser should be not worse than two diffraction limits at use of a telescope with diameter of main mirror $D = 1$ m is admissible.

References

1. V.V. Apollonov, Yu.S. Vagin, Patent № 2009118874, (2009)
2. V.V. Apollonov, A.J. Alcock, H.A. Baldis, Opt. Lett. **5**, 333 (1980)
3. V.V. Apollonov, V.V. Kijko, V.I. Kislov, Quantum Electron. **33**(9), 753 (2002)
4. C. Phipps, M. Michaelis, in *Conference Physics of Nuclear Induced Plasmas and Problems of Nuclear Pumped Lasers*, 26–30 Sept 1994
5. I.W. Campbell, Project ORION//NASA Technical Memorandum 108522, 1996
6. C.R. Phipps, J.R. Luke, D.J. Funk, D.S. Moore, J. Glownia, T. Lippert, Proc. SPIE **5448**, 1201 (2004)
7. V.V. Apollonov, Rep. Acad. Sci. (DAN) **351**(3), 339 (1996)
8. V.V. Apollonov, V.V. Kijko, V.V. Kislov, V.N. Tischenko, Proc. SPIE **5777**, 1011 (2005)
9. V.V. Apollonov, G.N. Grachev, A.I.Gulidov, Quantum electron. **34**(10), 941 (2004)

Chapter 17
High Power Lasers for New Applications

17.1 Lightnings

Lightning is the electrical spark discharge, which is manifested, usually, by the bright flash of light and by its accompanying thunder. Electrical nature of lightning was opened in studies of the American physicist B. Franklin, according to idea of which was carried out the experience on the extraction of electricity from the thunderstorm cloud. With this problem also dealt famous Russian scientists: M. Lomonosov and G. Rikhman. Lightning had been fixed besides the Earth on the surface of other planets: Venus, Jupiter, Saturn and Uranium. The average length of lightning 2–5 km, some discharges stretch in the atmosphere up to the distance to 150 km. Let us pause more in detail at the process of the appearance of lightning. Most frequently the lightning appears in the rain clouds, due to that they are called thunderstorm. Sometimes lightning can be formed in the layered—rain clouds, and also with the volcanic eruptions, the tornado and the dust storms. Usually are observed the linear lightning, which relate to the so-called electrode-less discharges, since they begin and end in the accumulations of the charged particles. This determines them some, until now, to not so clearly explained properties, which distinguish lightning from the discharges between the electrodes. Thus, lightning do not occur shorter than several 100 m; they appear in the electric fields considerably of weaker than field with the inter-electrode discharges; the collection of the charges, transferred by lightning, occurs for thousands of a second from huge number of small and well isolated from each other particles, located in the volume of several km^3. The process of the development of lightning in the thunderstorm clouds is most studied, in this case the lightning can pass to clouds themselves—intra-cloud lightning, and they can strike into the earth—ground-based lightning. For the appearance of lightning it is necessary that in the relatively small, but not less than the certain critical, the volume of cloud was formed the electric field with the tension, sufficient for the beginning of the electrical discharge ~ of 1 MV/m, and in the substantial part of the cloud there would be a field with the average tension, sufficient for maintaining the discharge ~ of 0.1–0.2 MV/m. In the lightning the electrical energy of cloud

V.V. Apollonov, *High-Power Optics*, Springer Series in Optical Sciences 192,
DOI 10.1007/978-3-319-10753-0_17

is converted into the thermal and the light. The process of the development of ground-based lightning consists of several stages. At the first stage in the zone, where electric field reaches critical value, begins the impact ionization, created by at first free electrons, always existing in a small quantity in air, which under the action of electric field acquire significant speeds in the direction to the earth and, colliding with the molecules, which compose air, they ionize them. Thus, appear the electron avalanches, which pass in the thread of electrical discharges—streamers, which are the well conducting channels, which, merging, give beginning to the bright thermo-ionized channel with the high conductivity—to stepped leader of lightning. The motion of leader to the earth's surface occurs by steps into several 10 m with a speed of ~50,000,000 m in second, after which its motion stops by several 10 μs, and glow strongly weakens; then in the subsequent stage leader again moves by several 10 m. Bright glow covers in this case all passed steps; then follow again stoppage and weakening of glow. These processes are repeated during the motion of leader to the earth's surface with a average speed of 200,000 m/s. The field strength at his end is strengthened in proportion to the advance of leader to the earth also under its action from the objects protruding on the earth's surface the reciprocal streamer, which is connected with the leader, is ejected. This special feature of lightning is used for creating the lightning rod. In the final stage on ionized by leader channel follow reverse (from bottom to top), or main thing, discharge of lightning, which is characterized by currents from ten to hundreds of thousands of amperes, by the brightness, which noticeably exceeds the brightness of leader, and with the high speed of advance, which at first reaches to ~100,000 km/s, while in the end of that being decreasing to ~10,000 km/s. The temperature of channel with the main discharge can exceed 25,000 °C. The length of the channel of lightning can be from 1 to 10 km, diameter—several centimeters. After pulse advancing of ionization current of channel and its glow are weakened. In the final stage the current of lightning can last the hundredth and even tenths it flogged, reaching hundred and thousands of amperes. Such lightning are called protracted, they most frequently cause fires. Main discharge discharges the part of the cloud frequently only. The charges, located on high altitudes, can give beginning to new arrow-shaped leader, who moves continuously with a speed of thousands of kilometers in second. The brightness of its glow is close to the brightness of stepped leader. When arrow-shaped leader reaches the earth's surface, follows the second main attack, similar to the first. Usually lightning includes several repeated discharges, but their number can reach also several ten. The duration of repeated lightning can exceed 1 s. The displacement of the channel of repeated lightning by wind creates the so-called strip lightning—luminous strip. With the entry of lightning directly into the soil is possible the formation of the unique mineral of fulgurite, which is, in essence, the sintered quartz sand. Intra-cloud lightning include usually only leader stages; their length varies from 1 to 150 km. The portion of intra-cloud lightning grows in proportion to displacement to the equator, changing from 0.5 in the temperate latitudes to 0.9 in the equatorial strip. The passage of lightning is accompanied by changes in the electrical and magnetic pour on and by the radio emission, by the so-called atmospheric disturbances.

The kill probability by the lightning of ground-based object grows in proportion to an increase in its height and with an increase in the electrical conductivity of soil on the surface or at a certain depth. The action of lightning conductor is based on these factors. If in the cloud there is an electric field, sufficient for maintaining the discharge, but insufficient for its appearance, the role of the initiator of lightning can carry out long wire cable, products of the combustion of the engine of aircraft or rocket. Specifically, thus sometimes "provoke" lightning in the layered—rain and cumulus congestus clouds.

For the illustration of enormous power engineering of the lightning and the significant frequency of their repetition on the terrestrial globe let us give article from the British press about "Marakaibo's lighthouse" (Venezuela).

There are more than three million discharges of lightning i.e. more than thousand for each second occurs each day in the world. The greatest frequency of discharges occurs in the tropical and subtropical zones of the terrestrial globe, the enumeration of these record places following: Ale hook in Columbia (270 days in the year with the lightning); Tororo in Uganda (274 days); Bogor in Java in Indonesia (283 days). Completely extraordinary light storm occurs in Venezuela in the mouth of Katatumbo river, that falls into Marakaibo Lake. This spectacle can be seen at night during more then 10 h of 180 days of the year. Light show occurs exclusively in the complete silence because of the intercloud discharges nature at the high up to 10 km altitude. Sky is illuminated due to that bright flashes of approximately 300 times an hour, that comprises more than one million electrical discharges in the year with the current strengths from 100 to 400 thousand ampere each. These discharges are visible from the ocean for many tens of kilometers and during many centuries were used by seafarers as the natural lighthouse, which was called "The Lighthouse of Marakaibo". Mixing cold wind from the Andes with the hot and humid air, which is risen from the swampy earth's surface in the environments Marakaibo is the major factor, which creates this unique phenomenon. Air being risen from the surface of swamps in essence consists of methane, which appears due to rotting of plants. Each flash of lightning could illuminate all lamps of South America. Now all depends on humanity which must find the method of the domestication of this enormous energy.

17.2 Lightning in the Natural Capacitor "Earth–Cloud"

Why does thunder and it does sparkle lightning? In the clouds are accumulated the electric charges, which leads to the breakdown of air, which is accompanied by the flash of light and by the formation of shock acoustic wave. By other words, thunder and lightning are the manifestation of atmospheric electricity existence. The fundamental connection of thunderstorm phenomena with the electricity for two and one-half of century was reliably confirmed by numerous studies of geophysicists, meteorologists, specialists on lightning protection. However, qualitative jump in the improvement of the methods of observations in the recent two decades led to the unexpected discoveries in the study of thunderstorm clouds, lightning, orbital scale of atmospheric electrical chain. It is similar, that the splash of interest in the problems of atmospheric electricity survived now will make it necessary to glance differently at the mechanisms of shaping of electric field and its role in the dynamics of atmosphere and ionosphere of the Earth. The heart of atmospheric electrical machine—thunderstorm cloud, is more precise, the totality of simultaneously "working" thousands of thunderstorms, distributed in the lower part of the atmosphere—to troposphere. Thunderstorm cloud lives not too for long—from the hour to several hours. But for the change to some thunderstorms come others, forming in the troposphere on the proximity. Contemporary satellite measurements, and also ground-based systems of the registration of lightning give to researchers the sufficiently reliable maps of the allocation of frequencies of lightning flashes over the Earth's surface. Flash rate above the surface of ocean on the average to the order is lower than above the continents in the tropics. The reason for this asymmetry is the intensive convection in the continental regions, where the land effectively heated by solar radiation. The quick ascent of thoroughly heated saturated by moisture air contributes to the formation of powerful convective clouds of vertical development, in upper part of which the temperature lower than ~40 °C. In the result are formed the particles of ice, soft hail, hail, whose interaction against the background of the rapid ascending current leads to the separation of charges. Cloud height in case of big water surface in average lower than above the continents, and the processes of electrization is less effective above the ocean. In lately is discussed another factor—difference in the concentrations of the aerosols above the ocean and the continents. Since aerosols serve as the condensation nuclei, necessary for forming the particles in the supercooled air, their abundance above the land increases the probability of the strong electrization of cloud. The quantitative analysis of this factor requires detailed experiments. Approximately 78 % of all lightning are recorded between 30 °S and 30 °N. The maximum value of the average density of the number of flashes per unit of the earth's surface exceeds 80 1/km^2 is observed in Africa (Ruanda). Entire pond of Kongo river with area about 3 million km^2 it regularly demonstrates the greatest lightning activity, although other equatorial centers abound with thunderstorms. It seemed that the significant contribution to the global thunderstorm activity the centers of thunderstorm of subtropics and middle latitudes give. Some of them, for example in North Argentina and Paraguay, lead on the rate of lightning flashes. The study of the regional special features of statistics requires further improvement of the satellite and ground environment of mapping thunderstorms. These studies intensively are conducted in the connection with the realization of lightning activity as the important factor of the climate of planet, which sensitively reacts to changes in the temperature, humidity, radiation conditions and the composition of the atmosphere. The progress in studies of atmospheric electricity is connected, first of all, with the study of the mechanisms of generation and dissipation of electrical energy in the atmosphere. Main problem here—the so-called electrical dynamo: the generation of quasi-stationary electric field and space charge in the moving weakly conductive medium. Already in simplest formulation of the problem of dynamo is manifested its important special feature—very wide interval of time-spatial scales. So that on the scale in several kilometers would arise the electric field, compared in the value with the breakdown of ~30 kV/cm for dry air under the standard conditions, necessary that the random traffic of charges with the collisions of cloud solid or liquid particles would lead to the matched effect of the addition of micro-currents into the macroscopic current of the completely high value (several amperes), sufficient for the rapid (tens of seconds) process of the separation of charges. It is evident from the experience that in the ripe thunderstorm cell the lightning the discharges occur with the enviable regularity—each 15–20 s, i.e. the mechanism of charging acting in the cloud is very effective, although the average electric charge density rarely exceeds several nCl/m^3. As showed the measurements of electric field on the earth's surface, and also inside the cloud medium (on the balloons, the aircraft and the rockets), in the typical thunderstorm cloud "main" negative charge—on the average several ten coulomb—occupies the interval of heights, which corresponds to temperatures from –10 to –25 °C. "Basic" positive charge composes also several ten coulomb, but it is located above basic negative; therefore the large part of the lightning discharges the cloud-earth returns to the earth negative charge. In the lower part of the cloud frequently is revealed smaller in the value the positive charge of ~10 Cl. For explaining the 3-layer structure of field and charge in thunderstorm cloud described above is examined the set of the mechanisms of the separation of charges. They depend on such factors as temperature, phase composition of medium, spectrum of the sizes of cloud particles. The dependence of the value of the charge from the electric field transferred for one collision is very important. From this parameter it is accepted to subdivide all mechanisms into the induction and the non-inductive. For the first class of mechanisms the charge q depends on value and direction of external electric field and is connected with the polarization of the interacting particles. Non-inductive of the charge exchanges between the being encountered particles in the explicit form on the field strength does not depend. In spite of the abundance of different microphysical mechanisms of electrization, now many authors consider the main thing non-inductive of charge exchanges with the collisions of small with the sizes from ones to tens of micrometers of the crystals of ice and particles of soft hail with the dimensions of the order of several millimeters. In the experiments was established the presence of the characteristic value of a temperature, at which the sign of charge q reverses, points of reverse, which lies usually between –15 and –20 °C. Specifically, this special feature made this mechanism so popular, since taking into account the typical temperature profile in the cloud it explains the 3-layer structure of the distribution of the charge density. Recent experiments showed, however that many thunderstorm clouds possess an even more complex structure of space charge. Is especially interesting mesoscale, which means the horizontal scales from ten to hundreds of kilometers, the convective systems, which serve as the important source of thunderstorm activity. Their characteristic feature is the presence of the united electrical structure, which includes the region of intensive convection and extended, sometimes to several 100 km the stratified region. In the region of stratification the ascending currents sufficiently weak, but the electric field has steady multilayer structure. Near the zero isotherm here are formed sufficiently narrow, with thickness into several 100 m, and the stable layers of space charge, mainly critical for the high lightning activity of mesoscale convective systems. A question about mechanism and laws governing the formation of the layer of positive charge in the environment of the zero isotherm remains debatable. As other examples of the work of electrical dynamo serve electrization and discharge phenomena with the dust and snowy storms, with the volcanic eruptions, nuclear explosions in the atmosphere, in the technological processes, connected with mixing of aerosol flows, for example in the flour-grinding and cement industry. The presence of powerful convective and turbulent flows in the thunderstorm clouds, and also the given above cases of electrization indicate that interaction of the large charged particles with the sizes of 0.1–10 mm with the particles of micrometer size in the moving conducting gas medium plays important role for the electrical dynamo. The poly-phase and multi-flow nature of medium brings in the final sum to the effective transformation of mechanical energy into the electrical. The problem of the initiation of lightning discharge remains one of sharp and unites entire complex of questions.

Briefly let us pause at two of them. First, as measurements on the balloons, the rockets and the aircraft showed, the maximum tension of electric field in the thunderstorm clouds does not exceed usually 2 kV/cm, which is substantially lower than the threshold of the breakdown of dry air at the heights in question, the composing value order 10 kV/cm. In the thunderstorm cloud, however, the discharge is developed in the humid air, which contains the particles of different size. Natural to assume that the threshold the field is reached in the compact spatial domain, for example, on the scale less than 10 m or generally on the microscale because of field strengthening on the particles, and to the sufficiently short period, so that the usual sensors of field do not ensure that required three-dimensional—time resolution with the registration. The development of the avalanche of the fast, so-called running away electrons with energy on the order of 1 MeV can serve as the alternative explanation to initiation of spark in the sufficiently weak field. Such electrons can be accelerated in the field of cloud because of a drop in brake power with an increase in the energy of particle. The threshold of the development of avalanche almost is by an order lower than the usual threshold of the breakdown of dry air; therefore when the seeding particles of the high energy are present, which are supplied by cosmic rays, the development of avalanche in the cloud can give birth to the localized region of high conductivity, capable of initiating spark. In the second place, the classical theory of gas discharge does not make it possible to explain the rapid passage of cloud medium into the conducting state at the preliminary stage of lightning. Recently the new scenario of this stage had been proposed and investigated, it is connected with the reaching by the cloud of the regime of the self-organized criticality. In the model of electrical cells with the significant dimension of ~1–30 m and with that randomly growing in the space and the time by potential the separate small-scale breakdown between the pair of cells is capable to cause "chain reaction" of intra-cloud micro-discharges—is developed the stochastic process "of the metallization" of intra-cloud medium. This model describes well the experimentally observed special features of the preliminary stage of the lightning discharge, in that number dynamics of micro-discharges and characteristic of radio emission. The study of thunderstorm phenomena includes experiments on the artificial initiation of lightning—trigger lightning. In order to cause the discharge in the atmosphere under the thunderstorm cloud, the rocket, which leaves after itself in the regime of a deficiency in the oxidizer a significant quantity of soot conducting electricity is used. As it proved to be, initiation occurs with the sufficiently high tension of field on the earth's surface under the cloud—not less than 5–10 kV/m. The for the first time classical diagram of the initiation of discharges with the wire pulling after the rocket was realized in 1960 from onboard of research vessel. From those times about thousands of successful experiments are carried out; physics of the ascending and arrow-shaped leaders, the return shock, mechanism of the sudden strengthening of the luminosity of lightning channel because of the sharp current amplification was in detail investigated because of them. Today are opened new possibilities for control of lightning discharges, in that number with the aid of lasers. Thus, powerful lasers are capable of organizing the extensive plasma cylinders in air, which could initiate and direct the lightning discharges similar to metallic wire in the trigger lightning and conducting channels from the products of the combustion of special geodetic rockets. Ch.Wilson has indicated in 1925 to the possibility of accelerating the electrons to the relativistic energies in the field of thunderstorm cloud. This hypothesis was confirmed during our days: on the generation of relativistic particles and quanta of high-frequency emission in the atmosphere, which correlate with the thunderstorm activity, a whole series of experiments testifies. Thus, measurements on the balloons revealed an anomalous increase in the intensity of X-ray and gamma-radiation more than to two or three orders in the course approximately of minute with the maximum of spectrum in the region 50–60 keV. Ground observations also revealed the intensive flashes of X-radiation with the energy of quanta into several 100 keV, whose appearance coincided with molding of the leader of lightning flash and the arrow-shaped leader of trigger lightning. Finally, for the latter several years were published the results of observations from the satellites of the splashes of the gamma-quanta MeV-foot level, X-ray and ultraviolet radiation of atmospheric origin. For explaining these phenomena in a number of works are drawn the phenomena mentioned above new for the atmospheric electrodynamics, and precisely: the generation of the running away electrons and breakdown on the running away electrons. Energetic quanta can arise from of the bremsstrahlung of the fast electrons, which interact with the molecules of air. Models describe different situations, including the development of the avalanche of fast electrons when the extensive air shower of cosmic rays is present, and breakdown on the running away electrons in the strong field of the stepped leader of lightning. One should note that an increase in the number of fast electrons in the field of thunderstorm cloud with the presence of extensive air shower is accompanied by the generation of a large quantity of second particles, and this leads to the generation of the current pulses and radio emission. If energy of primary particle is sufficiently great (10^{17}–10^{19} eV), the short (several microseconds) pulse of radio emission can have the enormous energy (to 1 MeV), which explains the appearance of the so-called narrow bipolar pulses, observed sometimes during ground-based and satellite measurements and correlating with the thunderstorm activity. The formation of such intensive current pulses is of interest both for understanding of the mechanism of the generation of lightning and for studying the cosmic rays of super high energies. It is interesting that the presence of the cellular structure of electric field in the thunderstorm cloud with the amplitude, which exceeds the threshold "of running away", proves to be essential for the process of accelerating the electrons to the relativistic energies. The randomly oriented electrical cells together with the acceleration sharply increase the time of life of relativistic electrons in the cloud because of the diffusion nature of their trajectories. This makes it possible to explain the significant duration of the splashes of X-ray and gamma-radiation and the nature of their interrelation with the lightning flashes. The role of cosmic rays for the atmospheric electricity they must explain experiments on a study of their correlation with the studied thunderstorm phenomena.

17.3 Lightning in the Natural Capacitor "Cloud–Ionosphere"

The powerful pulse to the study of atmospheric electricity at the very end XX of century gave observations of the discharge phenomena in the middle atmosphere—Sprites and Jets (Figs. 17.1 and 17.2), which correlate with the thunderstorm activity, each day where the region of glow stretches to 85–90 km above the earth, and flash duration comprises from ones to tens of milliseconds, elves,

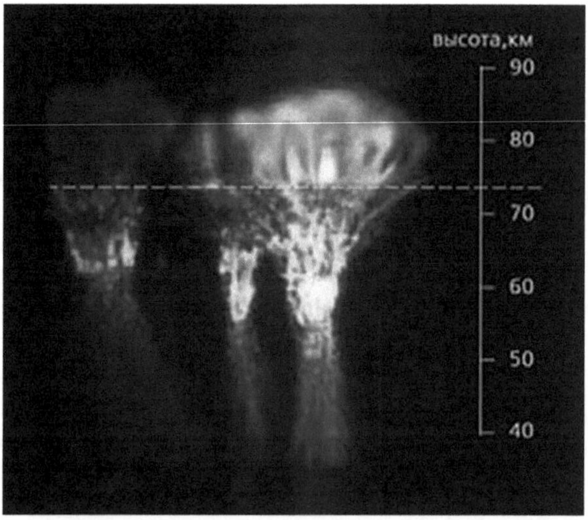

Fig. 17.1 Photo of "Sprite"

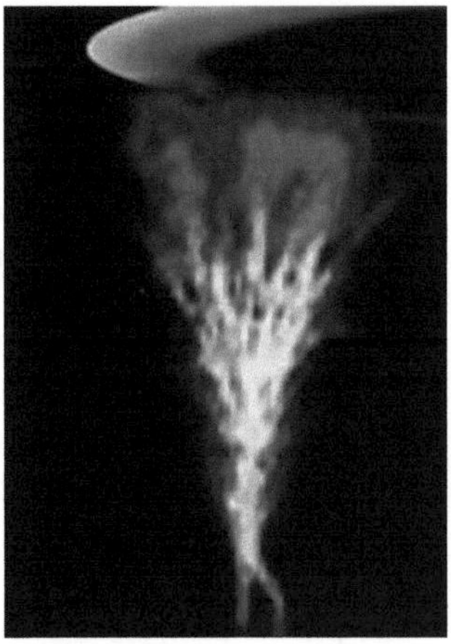

Fig. 17.2 Photo of "Jet"

who stretch to the heights—70–90 km with the duration less than 100 s and jet—discharges, which start from the upper part of the cloud and which are extended at times to the mesospheric heights with a speed of about 100 km/s. Registration of high-altitude discharges and diagnostics of their characteristics are complicated because of the short time of life and sporadic nature of the lightning discharges. Partly therefore, until now the physical models of Sprites, Elves and Jets remain the object of intensive discussions. Perhaps, the greatest information to the present time is accumulated about Sprite. Jet and its physical examination is the most interesting stage of the work for the future.

The optical flash of Sprite in the mesosphere appears through several milliseconds after discharge cloud (+)—the earth, sometimes at the removal several 10 km along horizontal from the channel of lightning. Discharge ignites at a sufficiently high altitude, since the threshold of the breakdown of air falls with the height exponentially, whereas the amplitude of the disturbances of electric field, which appear at the moment of the powerful lightning flashes of capacitor "cloud—the earth", decreases with the height much slower—according to the power law, and at the heights approximately 75 km in the atmosphere exceeds the threshold of breakdown. In the last decade the study of the high-altitude discharges in the atmosphere became the extensive, intensively developing direction of geophysical electrodynamics. And although the stage of the accumulation of data, that characterize the morphology of these phenomena, is by no means completed, already it is possible to switch over to a study of the thinner special features of structure and dynamics of the high-altitude discharges and their role in the orbital electrical chain and the balance of low components of the atmosphere. Experiments and theoretical calculations show that these discharges must be considered as the part of the dynamic process, which includes the formation of the structure of field and charge in the cloud and lightning discharge to the earth. Widely are discussed different approaches to the simulation, including with the attraction of the mechanism of the running away electrons. In the detail, was developed the model of the generation of electric field in the middle atmosphere, which considers the special features of the charge distribution and its dynamics in orbital scale the convective systems, which serve as the basic sources of positive discharges to the earth. The physical model had been developed, which makes it possible to describe fine structure and dynamics of Sprite. Sprite is represented as the network of micro-discharges—streamers—and as the self-sustaining process in the external field. Together with metallizing process of thunderstorm cloud mentioned above, we have here one additional example to self-organizing, when the dynamics of high-altitude discharge is caused by reaching the threshold of the so-called directed flow, which characterizes the formation of the branched conducting channels, which overlap entire length of Sprite. Immediate prospects in the study of the high-altitude discharges in the atmosphere are connected with conducting of the coordinated ground-based and space experiments, and also with the development of the new methods of diagnostics of the parameters of high-altitude discharges, in that number with the use of possibilities of radio-wave and laser diagnostics. Electrical field of the atmosphere is very variable. The tension of the vertical component of field, which usually much more than horizontal reaches several kV/m with taking into account the ground-winds and the thunderstorm cloudiness. Therefore is introduced the concept of conditions grow prettier weather, that correspond to wind speed not more than 6 m/s into the absence of any kind of sediments, hoarfrost, fog, lower cloudiness. But even under these conditions near the earth's surface there is an electric field by tension approximately 150 V/m, along which in the weakly conducting air flows the electric current with the density several pa/m^2. This field changes in the time and the space, the fluctuations relative to average value can comprise from ones to tens of percent. The measurements of electric field, current and conductivity under the conditions grow prettier weather they serve as the powerful means of the study of the electrical state of the atmosphere. However, to use it is possible, having only learned to divide global i.e. planetary scale, orbital—with the scale of the order of the height of neutral atmosphere 100 km and the local disturbances of the electrical parameters. The latter, in their turn, are directly connected both with the changes in the ionic-aerosol composition and with the dynamics of medium. In studies it was established that the boundary layer of the atmosphere is characterized by the presence of the aeroelectrical structures, which are manifested in the pulsations of electric field with the power-law spectra short-term with the periods from ones to several 100 s. The electrodynamics of fog is of special interest. Full-scale observations showed that under the conditions of fog in the frequency band 10^{-2}–1 Hz also realize the power-law spectra of the pulsations of field, but the intensity of pulsations increases more than by the order in the comparison with the conditions grow prettier weather, which is explained by the diffusion charging of drops in the lowest layer of the atmosphere. The results it is structural—temporary analysis they made it possible to isolate two forms of the electrical state of fog, the first of which is characterized by the formation of intensive aeroelectrical structures, and the second—by chaotic structural—temporary variations in the field and current.

17.4 Orbital Electrical Socket

Electricity grow prettier weather is inseparably connected with the thunderstorm electricity it composes the part of the distributed current outline—the orbital electrical socket (OES) of the atmosphere. As the physical cause for formation OES in the atmosphere serves a sharp increase in the conductivity of air with the height. Near the earth's surface the conductivity of air is very small and comprises (2–3) × 10^{-14} S/m, which corresponds to the concentration of light ions of approximately 10^3 cm^{-3}. With an increase in the height because of an increase in the ionization level, determined to 40 km by galactic cosmic rays, and it is above—by the ultraviolet and X-radiation of the sun, conductivity grows almost exponentially with the characteristic scale of 6 km. Already at the height of lower boundary of the ionosphere (about 80 km) it increases to more than 10 orders in the comparison with the troposphere. The conductivity of the earth in the surface layer, and all the more water in the ocean also exceeds the conductivity of the boundary layer of the atmosphere to 10–12 orders. Thus, the constantly functioning thunderstorm generators prove to be concentrated in the sufficiently narrow weakly conducting layer between the earth's surface and the ionosphere. Frequently with the simplified description OES the earth's surface and lower boundary the ionosphere (70–80 km) they are considered as the facings one additional gigantic spherical capacitor of the orbital scale, which is discharged in the regions grow prettier weather it is charged in the regions of thunderstorm activity. In this case the quasi-stationary currents of charging are not locked completely to the earth near the thunderstorm clouds, but partially "they are involved" into the superincumbent region of high conductivity and spread on the ionosphere. It is considered that the precisely quasi-stationary currents in the first turn "bear responsibility" for the maintenance of a potential difference of ~350 kV between the Ionosphere and the Earth. Since the upper part of the majority of thunderstorm clouds has positive charge, the potential of the ionosphere also proves to be positive, and in the regions grow prettier weather electric field directed downward, causing thus the conduction currents, which close OES. If the action of generators ceased, a potential difference between the earth's surface and the ionosphere would disappear in the time not more than 10 min. The orbital electric generator, caused by the nonmonolithic nature of the rotation of the plasma sheath of planet, can serve together with the thunderstorm generators, the potentially important source of atmospheric electricity. In order to understand in the components the mechanism of work OES and its role in the system solarly—terrestrial connections, to be dismantled at is three-dimensional—to temporary dynamics, are necessary the complex experiments, which include the high-precision measurements of electric field and current on the orbital scales and the measurements of the integral activity of thunderstorm generators. For understanding of work OES are necessary the adequate models of thunderstorm generators, which give the possibility to calculate quasi-stationary and pulse currents, and also the contribution of generators to complete power engineering OES. The results of calculations show that the contribution of mesoscale convective systems in OES can substantially exceed the contribution of single thunderstorms. In spite of the significant disturbance of electric potential in the region of convection, which reaches hundred megavolts, the basic contribution to the complete vertical current and structure pour on and currents in the environment of mesoscale convective systems are determined by region stratifications. Taking into account last experimental data and based on them theoretical estimations was created new description OES as the open dissipation system, supported in the state of dynamic equilibrium by the entering it energy flow, in the first turn of radiant energy of the sun. It is established that electrical energy is generated predominantly in the regions of the reduced atmospheric pressure and in the zones of cold fronts and comprises in average 5×10^{13} George, that two orders higher than energy, concentrated in the orbital spherical capacitor the earth-ionosphere. The average speed of the dissipation of electrical energy 4×10^{11} W exceeds the speed of the dissipation of energy in the tectonic and magmatic processes and is characterized OES as completely dynamic system with the very short time of the renovation of electrical energy. The process of the transformation of energy in the atmosphere is accompanied by the generation of the dissipative structures of different scales, in the detail of convective systems and aeroelectrical structures, that clearly it is possible to visualize in the form the global processes of the accumulation of electrical energy and its dissipation in the atmosphere. On the basis of higher than aforesaid, in the near time the study of power engineering of the atmosphere, structure and dynamics OES will remain one of the vital problems of atmospheric electricity.

17.5 Space Flights by Laser Light

LJE is substantially more economical than traditional chemical-propellant engines. In the initial stage of flight as the working medium adapts atmospheric air, and beyond the limits of the atmosphere the insignificant onboard reserve of gas or easily sublimated substance. In this case specific expenditures for the launch of loads into space can be reduced to 5,000–10,000 rub/kg, which is approximately 100 times less in the comparison with the contemporary scale of world prices. Especially, the possibility of maintaining the orbital parameters at the assigned level with the aid of the same laser system, intended for the starting, can be evaluated very highly.

In the present time the set of works on the study of LJE designing possibility is under very intensive realization. Within the framework project "Lightcraft" the intensive developments of such systems are conducted in USA, Germany and Japan. So, in November 2,000 year American company "Lightcraft Technologies" successfully conducted model tests of rocket, which rose to the height –70 m during 12.7 s under the action of exhaust jet, which appears under the effect of powerful laser emission. In the experiments the low-frequency 20 Hz CO_2-laser with output power 10 kW was used, which in the present time is under upgrading to the level of the output power—100 kW. Mechanical impulse appeared due to ablation of special polymeric material from the concave surface of the reflector, located in the tail end of the rocket, where the laser beam pulses were accepted.

In the beginning of 1973 in the USSR under scientific guidance of Acad. A. M. Prokhorov the work on the study of possibility of LJE designing was conducted. The reflector, located in the tail of the rocket prototype, concentrated the obtained radiation in air and ensured micro-burst that the reactive thrust was created. The successful results of different models of the reflectors tests, which were at the same time the laser light receivers, had been obtained. One should note that all experiments pointed out above were conducted with the use of electric discharge CO_2-lasers with power up to 10 kW, while for the injection into orbit of different highly and technologically effective equipment (global network connections, Internet, photo-monitoring of Earth surface, debris cleaning) the radiated power substantially higher is required Thus, for example, for KA launching with the weight 1,000 kg the laser with power not less than 10–15 MW is necessary. This laser at present time can be the gas-dynamic laser (GDL) only, since only in this case the laser technology in the significant measure intersects the rocket technology, which in 50 years moved sufficiently well in the creation of super-power gas generators and it makes it possible to pose similar problems. Furthermore, the laser must work in P–P regime with the high frequency of the repetition of short pulses for the exception of the process of laser emission screening entering by the plasma, which appears with the engine work, and so for an increase in the work effectiveness [1]. In the opinion of specialists—classical missileman—LJE can find an effective use in creation of cheap single-stage means of nano- and micro-satellites launch with mass in the gap 5–50 kg, which according to the opinion of experts tomorrow will compose the fundamental basis of commercial launch program. In the first stage of flight KA at the heights up to 30 km as the working medium in the engine the usage of atmospheric air is intended, and then to inject satellite into orbit the onboard reserve of the special fuel—ablating substance- in the quantities not exceeding 15–20 % of the KA started weight is needed.

The experience of powerful lasers creation is accumulated in the A.M. Prokhorov General Physics Institute of RAS, in the NPO "Energomash" after V.P. Glushko and other organizations of Russia. In OOO "Energomashtechnika" during last few years very successful experimental studies of P–P regime in the powerful and well developed continuous laser systems were conducted. This makes it possible to approach the experimental realization of super-power P–P laser source on the basis of gas-dynamic principle and LJE in the composition of light carrier with control system. The forthcoming complex of works must become major step toward the future starting of super-lightweight KA into low near Earth orbit. The realization of project will make it possible to create highly economical LJE of reusable starting for the removal of the payloads of wide designation into space. The key advantage of new approach is connected with the fact that the source of kinetic energy and payload are untied in the space and launching weight KA can be lowered to the payload weight only. An even greater K.E. Tsiolkovskii foretold, that the starting KA of future will be achieved with the aid of the electromagnetic waves directed from the external energy source, laser at that time was not known to the World at all. The great interest of scientists and specialists, as they confirmed few last symposium in this field of research, was connected with the successful solution in our country of the problem of powerful lasers creation with high repetition rate (50–100 kHz) of pulses of short duration (150–250 ns). This is what very recently wrote "the father of laser reactive motion" Arthur Kantrowits:

> Dear Victor, thanks for the pleasure given to read your articles from the last symposium on the motion with the aid of laser energy which are now accessible for the scientific community. I already thought above your works and I count that these remarkable ideas on the quasi-stationary wave, about the light-detonation waves, about the matrix of reflectors those introduced into your new examination are very important for the development of theory and technology of motion with the aid of the light. With great admiration I do control the development of the laser starting in the world and in Russia in particular. I hope that after your publications the significant activity in the field of laser engine creation and its applications will be stimulated. It would be very interestingly and further hear about the progress in this hot field of research. Thanks for your energetic contribution and be healthy.
>
> A. Kantrowits.

High-frequency P–P mode of laser operation is developed and tested on the base of powerful CO_2—GDL, Nd YAG with lamp pumping and can be successfully used for other types of the powerful lasers, such as: HF/DF, Nd YAG with LD pumping, COIL and CO. The results of computer modeling for high repetition rate P–P lasers are presented in our review. The data of modeling are dependable on the characteristics of media, way of pumping, resonator geometry, and many other parameters of the laser under consideration [2].

1. COIL, P–P mode starts at frequencies >20 kHz, Depth of modulation—100 % starts at frequencies >100 kHz., Pulse duration <250 ns. Ratio: P peak./Paver. = 100–1000.
2. HF/DF, P–P mode starts at frequencies >100 kHz., Depth of modulation —100 % starts at frequencies >250 kHz., Pulse duration <150 ns. Ratio: P peak./Paver. = 1000–10000.
3. Nd YAG, P–P mode starts at frequencies >4 kHz., Depth of modulation—100 % starts at frequencies >40 kHz., Pulse duration <250 ns. Ratio: P peak./Paver. = 100–1000.
4. CO, P–P mode starts at frequencies >10 kHz., Depth of modulation—100 % starts at frequencies >100 kHz., Pulse duration <250 ns. Ratio: P peak./Paver. = 100–1000.

At present time the project of the GDL with an output power of 20 MW with the varied high-frequency structure of emission is located in the stage of active study in our country. Meaningful results on a notable increase in the effectiveness of laser energy use in the developed mode of emission, which made it possible according to our estimations to pass from tens of kg pay-load to a hundreds, or even a thousand kg are obtained. So, one should note that the works in this region, on the basis of their large prospect for different applications, already begun in

Germany, Japan, England, France, China, S. Korea, Brazil and a number of others the countries. In this case, practically all participants of that works note the chosen position of GDL as the most promising system from the point of view of its scalability on the basis of the rocket technology to the level of several 10 MW and other parameters, important for this kind of applications.

To the present time two directions, in which the possibility of applying the laser emission in the aerospace tasks is investigated: launching to the space orbit light KA, reduction of aerodynamic drag of the flying bodies, which move in the atmosphere with the high speed were formed. In the problem of developing LJE the generated by P–P laser emission is focused by reflector near the rear end of KA and the periodically repetitive laser sparks are created. Sparks generate shock waves, which transfer the part of their mechanical pulse to the reflector, located near the tail of the rocket. As the characteristic of the previous years the gas-discharge laser systems had demonstrated the pulse repetition frequency, which was limited by the time of gas exchange in the discharge zone in the range of 100–300 Hz. For achievement high average power—10 MW for this rate of operation it should be necessary to use laser pulses with the energy about 100 kJ. With the reduced air pressure, for example for the height more than 15 km, long living plasma sphere, created by each pulse, occupies practically entire volume of reflector, which leads to the screening of the subsequent pulses into the course ~ of 10 ms. Technical difficulty of that method of energy extraction for the laser is also a very strong impact loads with such a high energy of pulses. The use of high energy pulses with the small repetition frequency and, consequently, with the very high peak power is limited also by optical breakdown as on the route, so on the surface of reflector. The method proposed by us is the way of overcoming all that difficulties on the indicated basis: usage of laser emission with short duration of pulse and high pulse repetition frequency and mechanism of generated by OPD shock waves resonant association. Under that conditions of LJE energy of laser pulse with the short duration (100 ns) effectively (95 %) can be absorbed and converted (~30 %) into shock waves. In addition to this it is shown that the specific thrust can be increased several times due to the artificial transformation of radial component of shock waves into the longitudinal one. There are many other advantages of high repetition rate P–P lasers should be mentioned here as well [3]. They are:

1. Much higher efficiency of energy extraction from a media and conversion into laser light, which is important for high aperture lasers scale up;
2. Plasma screening effect decreasing due to the decreasing of laser pulses duration and increasing of number of pulses up to a few tens of kHz;
3. Decreasing of laser beam thermal defocusing due to the optimal temporal structure of laser radiation;
4. Localization of energy deposition in space and time and elimination of stressed state of solids under laser action due to replacement of melting by ablation;
5. Possibility of unrolled 3d geometry of breakdowns in space due to a very high repetition rate generation of laser pulses, creation of optical, acoustical and electromagnetic fields far away of laser source;
6. Much longer distances of optical breakdown conditions for the same optical systems.

17.6 Spark Parameters for LJE

Laser emission is focused by the reflector, which can have a form of hemisphere or parabolic. In Fig. 17.1 the significant dimensions of reflector, focusing one spark, and also plasma zone, created by spark are shown. Distance F_f must be little between the focal point and the reflector $F_f/R_d < 0.2$, which follows from the condition of achieving the high value of the specific impulse of thrust. It follows in avoiding of the optical breakdown on the reflector and of the conditions of transport that the beam must have large diameter on the reflector. If the intensity of emission exceeds the threshold of optical breakdown, then plasma front is extended towards the ray, occurs heating and the ionization of air as a result of absorbing the laser emission. Since intensity in the sharply focused ray rapidly decreases (shape factor), then already at the small distance from the focus occurs the disruption of the light-detonation regime of the propagation of plasma front. Further, emission is absorbed for a while in the decay plasma. It follows from the calculations carried out by us that for LJE the greatest interest represents the pulses with the duration ~ of 0.2 s. Maximum energy of laser pulses is limited by the condition for the achievement of a high efficiency in the use of laser emission for the thrust application. In the free gas space the pulse transferred by shock wave is different from zero only at the small distances from the burst center. The same data were observed also in the early experiments. For the reflector of parabolic form the specific impulse is maximum ~550 N·s/J at the distance R_1, equal with the short duration of the pulse of ~100–200 ns, approximately, the tenth of a dynamic radius, or by another words, the distance, where the value of pulse diminishes is approximately three times bigger. With the increase R_1/R_d from ~0.1 to 0.33 the value J_1 it rapidly decreases ~ by 550–200 N·s/J. B the case LJE the possibility of using the low values R_1/R_d it is limited by the formation of the long-life plasma, whose characteristic radius is compared with the dimensions of reflector. At the last stages of the thermal expansion of spark is formed plasma zone with the low density and the high temperature ~8,000 k^0 the ionized gas. The contact boundary of hot region stops with the pressure balance in the plasma and the surrounding gas. In the approximation of the spherical form of spark a radius of plasma region at the pullout moment of pressure it is possible to obtain the estimation of a radius of plasma formation for entire operating range of the pressures 0.1–1 atm. The ratio of a radius of plasma zone to a dynamic radius in this case composes the value 0.15–0.25. This relation does not depend on energy and the gas pressure. Moreover its value is located in the same range, where is attained the maximum the value of the specific recoil momentum, created by shock wave. In the reflector, which has the form of hemisphere, the maximum the specific impulse of the thrust has reached at $R_1/R_d \approx 1$. Let us give the values of a radius of plasma formation and time of its formation for energy 10^5 J and the pressures of atmosphere 1 and 0.1 atm. The parameters correspond: peak pulse power—2×10^7 W and to the pulse repetition frequency—100 Hz, and also to start and to the end of the stage of acceleration KA in the regime LJE. With pressure 1 atm. the size

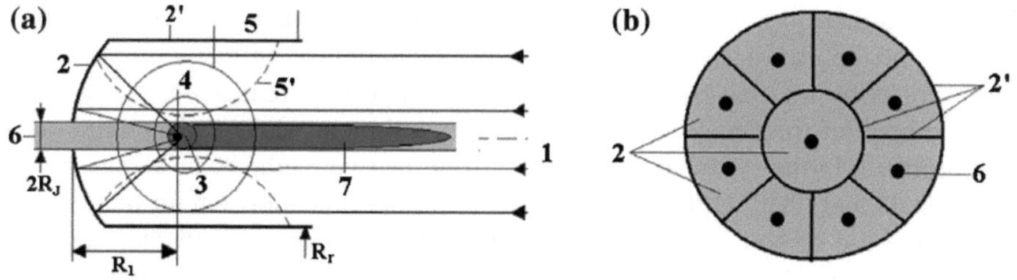

Fig. 17.3 Scheme of reflector (**a**) and possible structure of matrix of reflectors (**b**) for LJE. *1* high repetition rate P–P laser radiation, *2* back side of reflector-zone of laser pulses interaction, *2′* side wall of reflector, *3* cavern, *4* OPD, *5* Shock wave, *5′* reflected shock wave, *6* gas stream, *7* plasma stream

of plasma sphere will be 25 cm. With the time of formation on the order of 1 ms, while at a pressure of the atmosphere—0.1 atm. the size of plasma sphere will comprise already 50 cm with the time of formation—2 ms. Cooling laser plasma occurs due to the turbulent mixing with the cold surrounding gas. The characteristic time of this process more than to the order exceeds the time of its formation. Here the spark with spherical form has been examined. On the matter itself the spark in LJE has a form of cone with the large apex angle. This even more worsens situation. With the specified conditions at the last stage of the expansion of spark can be formed the cumulative jet, in which the gas moves in the direction to the reflector. Plasma ceases to be singly connected, taking the form, similar to the torus. Thus, the use of a regime with the maximum pulse, presented in the early work of many authors, it will be accompanied by the contact of plasma with the surface of reflector, which, undoubtedly, will lead to its destruction (Fig. 17.3).

17.7 Mechanism of Shock Waves Resonance Association

The effect of resonance association separate shock waves into the low-frequency quasi-stationary wave (QSW) in the general case consists in following. In continuous medium consecutively are created periodic disturbances—shock waves, whose initial velocity is more than the speed of sound. If the speed of the displacement of the field of pulsations is lower than the speed of sound in the medium, then shock waves are united and are created BY SQ. Depending on the time-spatial structure of pulsations the mechanism is manifested in the form the effects, basic property of which is the large length of high-pressure area. The mechanism of shock waves association does not contain limitations to the type of medium and source of pulsations, to its energy. Depending on the structure of pulsations QSW can have various forms.

The realization of mechanism kV makes it possible to remove the problems examined above: the screening of emission and the thermal action of laser plasma in LJE. Furthermore, QSW makes it possible to substantially increase the effectiveness of the use of laser emission in the calculation of an increase in the specific thrusts by the unit of power. Let us determine requirements for energy and pulse repetition frequency of P–P of the laser emission, which satisfy solution of this problem. In this case has the sense to examine two methods, based on the use "spherical" and "flat" OPD. In both cases is formed flat short circuit the calculation of the use of both the geometry OPD and the geometry of reflector, and also the organization of the process of the introduction of energy of laser emission into the engine.

17.8 LJE on the Basis of Shock Waves Resonance Association for MR

In the general case MR is represented in the Fig. 17.1. The problem of developing LJE on the basis of shock waves merging and P–P of laser emission with the high (to 100 kHz) pulse repetition frequency, and also flight path control aid LJE it is decided with the use MR. The schematic of engine is collection from N monoreflectors. To MR brings by P–P the emission with the energy of the pulses q with the average power W_C. In the general case the elements MR are identical. MR creates matrix OPD, each of which is stabilized by the air jet of the incident flow. A quantity of matrix elements in the general case determines the high-speed regime of gas jet. In our case a quantity of elements in the matrix was $N = 8$.

In our calculation of power of laser emission was taken as the equal of 20 MW, the pulse repetition rate—100 kHz, energy of each pulse in this case was equal 200 J. Calculations showed that for the accepted for the examination parameters in the case of normal atmospheric pressure the complete engine thrust will be equal to 100 thousand N, and at a pressure of the order of the tenth atmosphere its value will fall to the value—36,000 N.

The optimum gas flow speed for both cases proved to be equal to 2.5 and 5.4 km/s respectively. One should also note that the flight control is possible with the aid of the laser engine for the calculation of a change in the thrust in the equivalent components MR. However, an increase in the number of elements MR is accompanied by a significant increase in aerodynamic drag on the air ducts and the selected number of elements is close to the optimum.

In the experiments the model of rocket with the laser engine was the cylinder with diameter ~8 cm, with the length ~26 cm and with the weight 1.1 kg, suspended to four thin wires with total length of 1.1 m (Fig. 17.4). Construction allowed displacement only in the axial direction. On the chamber end is fixed the reflector—change cylindrical cap. Laser emission was introduced into the camera through the lens with a focal length of 17 cm. The jet of argon was formed with the expiration from the high-pressure chamber through the opening with a diameter of ~3 or 4 mm. The speed was regulated by the pressure of argon, which was given into the camera on the flexible hose. The force, created by jet and shock

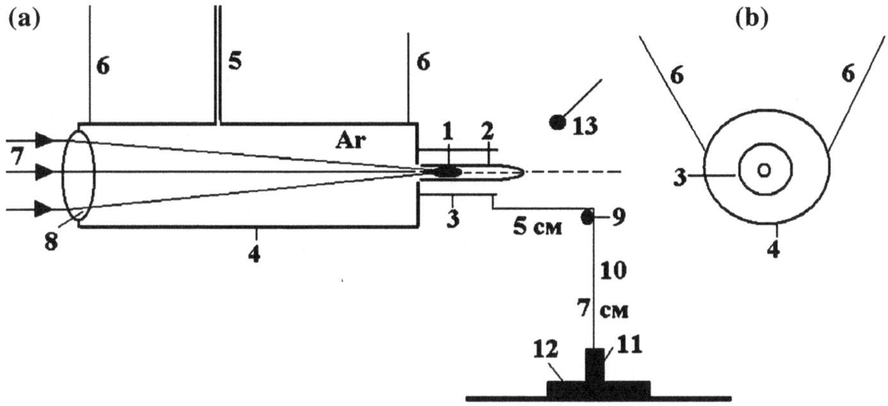

Fig. 17.4 Scheme of experimental setup

waves, was transferred with the aid of the thin (~0.2 mm) molybdenum wire to the load, which stands on the strain scale weights.

OPD burnt in the flow, which was formed with the outflow of gas from the camera through the opening in the reflector. OPD was created by P–P laser emission—f = 50 kHz and 100 kHz, in this case the average radiated power comprised P ≈ 1200 W. Thrusts with f = 50 kHz, The V = 300 m/c it proved to be equal—40 g, and with The V = 400 m/c – 69 g. The value of the thrust coupling coefficient has proved to be equal—1.06 N/kW. Experiment confirmed that P–P laser emission creates stationary thrust with the very high coupling coefficient—1000 N/MW. A difference between the experimental results and theoretical estimations, whose value >4500 N/MW, is connected with the need for further optimization of the process of the introduction of energy of laser into the plasma OPD.

17.9 Resonance Properties of System “Laser-KA”

By important parameter in the estimation of the resonance properties of system “Laser-KA” the ratio of the duration of laser pulse to the transit time of sound throughout the entire length of rocket appears. In the practical sense there is greatest interest in the case, when uniform load under the action of laser pulse is created throughout the entire length of rocket. If pulse has very short duration, then acceleration is non-stationary, the length of that excited in KA waves are much less than the length of rocket. The case, when the pulse duration and the transit time of acoustic wave are compared, corresponds to the resonant step-up of waves.

Let us make the resonance properties of system for the laboratory experiences in the case of KA with the mass—20 kg, the length of the rocket of equal—200 cm, acceleration—100 m/s^2. Average power of P–P radiation—4 MW, the energy of pulses generated with repetition rate—200 Hz is equal—20 kJ. Calculations show that for small pulse repetition rate of P–P laser radiation the acceleration mode is accompanied by the extremely strong dynamic structural loads KA. They have resonance nature, since the wavelength of compression it is compared with the length KA. Furthermore, with an increase in the length KA about 4 m and the repetition frequency of the laser pulses up to—1 kHz the natural frequency of oscillation of rocket is equal to pulse repetition rate of laser and we obtain the fulfillment of conditions of resonance. This is very dangerous, since resonance loads can exceed the safety factor of rocket and it will be destroyed in the flight.

Thus, estimations show that the thermal contact of plasma with the reflector and strong dynamic loads are inevitable with the low pulse repetition frequency. Situation is aggravated by the excitation of resonance oscillations in the KA. These difficulties are surmounted only with the use of a method, based on the association of shock waves [2]. Calculations and experiment confirmed possibility of significant stationary thrusts designing with the use P–P of emission with high pulse repetition frequency.

17.10 Lightning Protection System

From other side during the last 20 years in many laboratories of the World very interesting studies of effective laser lightning protection system creation had been conducted. On the basis of pulsed laser long enough laser spark can be created, which is able to connect thunderstorm cloud with the grounded iron core—classical lightning rod. The maximum lengths—16 m of such highly controlled by the laser electrical discharge channel—have been obtained in Russia and in Japan with help of developed and produced in A.M. Prokhorov GPI RAS pulsed CO_2 lasers with the energy up to 0.5 kJ.

Later also we have demonstrated much longer laser spark, obtained with the aid of conical optics, which has exposed much more advanced characteristics from the point of view of plasma channel conductivity and very short time delay of its effective start. In our works for the first time the possibility to use as the single wire connecting two resonant circuits by long laser spark, created with the help of conical optics, had been suggested.

At present time among an enormous quantity of different type of lasers for electrical discharges directed by the laser spark wide spectrum of application only two type of lasers are found as the most effective for this particular task: pulsed sub-microsecond CO_2 lasers and femtosecond (FS) solid state lasers.

Possibility of one or several parallel highly ionized channels creation, so called filaments, with a characteristic diameter of 100 mkm in the atmosphere along the direction of the laser beam propagation is the major feature of FS lasers. With estimated value of electron densities of such a filaments (up to 10^{16} $1/cm^{3)}$ and with wavelength of the laser in the interval 0.5–1.0 mkm plasma, practically, does not absorb laser emission. In this case the length of channel is determined by the intensity of laser emission and in the case of energy of the FS pulse ~ of 100 mJ can reach the length up to 150 m. The possibility of a considerable increase of filaments length is connected with the application of unique super powerful FS laser systems with the energy about 1 J and even more. Such a system in our days looks like a unique laboratory setup and hardly can be used for commercial and other applications. Another limiting factor is a very high resistance of conducting channel based on FS laser filaments. Characteristic value of resistance of such a channel is on the level of 0.1 Mohm/m [2, 4].

In the case of pulsed CO_2 laser use in combination with classical spherical or conical optics the length of spark channel is determined by the parameters the caustics of laser and that its part in which the intensity of emission it exceeds threshold for the optical breakdown of air. The presence of intensity threshold is connected with the special features of the mechanism of plasma formation where the avalanche multiplication of initial "background" electrons in the field of laser emission take place. A vital difference in the geometry of spark in the case of long pulse and spherical optics from the case with the FS lasers is the fact that the laser spark occupies volume inside the caustic curve from the focal plane to the side of laser, since the laser radiation cannot traverse self created plasma shield. In this case for the super long spark formation important to use CO_2 laser pulses with traditional form, i.e., with short (50–100 ns) spike and long (1–3 mks) tail [3]. Long tail ensures the additional heating of laser plasma, moving along with the laser beam by expanding to the caustic and, respectively, increasing the length of laser spark. The length of laser spark in several 100 m has been obtained even in the USSR with the aid CO_2 laser with output energy ~ of 5 kJ and the classical form of the pulse.

However, in the case of conical optics use the possibility of laser plasma channels formation of substantially larger length does appear. According to our estimations the length of such a channels in the case of conical optics in combination with pulse-periodic (P–P) CO_2 lasers with an average power of 0.1–1 MW can reach several kilometers. In this case the geometry of spark is completely different and the region "of collapse" is determined by the diameter of the beam on the surface of conical mirror only and by the angle of taper. Spherical optics plasma front moves in the direction of laser beam, i.e., from the surface of the focusing mirror in outside, in this case different spatial domains in the section of laser beam form different sections of plasma cylinder. Therefore the formation of very long laser sparks even with comparatively short lengths of laser pulses is possible. The photo registration of spark showed that it consists of the set of the almost contiguous between themselves elements, on the form resembling "Chinese cap" and located on the identical distance from each other.

Thus, for the experiments conducting the electrical energy transfer between two resonant circuits, situated on the distance not more than 100 m, it is possible to use both type of lasers. At the same time for commercial energy transmission system creation the powerful high repetition rate P–P CO_2 lasers appear more promising, since they possess the substantially lower threshold of the optical breakdown in air and technically are more advanced in scaling up of output power to many hundreds of kilowatts.

As has already been spoken above, the geometry of plasma cylinders for two types of laser was completely different: the collection of thin filaments-FS laser or cellular single channel with more or less uniform plasma—powerful sub-microsecond CO_2. At the present time it is possible to assert that powerful pulse CO_2 laser and the mechanism of laser radiation breakdown of gas medium have the decisive advantages over the FS ideology of the energy transfer by long lengths laser plasma channel.

First experiments carried out by our group with pulsed CO_2 laser having the traditional form of lasing in the beginning of 90th were very much informative. By potential applied to "the ends" of laser spark (up to 600 V) we have showed that the glow of laser plasma remains not more than 10 mks, at the same time the conductivity of laser plasma channel does exist considerably longer—up to 100 mks. High repetition rate P–P laser emission is capable to solve the tasks of the reproduction of the conducting channel during long period of time and to increase the conducting channel length significantly. The most essential drawback of the described method is the fact that it cannot be directly used in vacuum, which means beyond the limits of the Earth's atmosphere.

New method of the energy transfer and new approach to the transmitting channel formation are directed toward essential simplification of electrical energy transfer methods, at the same time they can be effectively applied to the non-conducting media conditions. Necessary result can be achieved due to the "Impulsar" technology.

The energy transfer in this case is achieved because of the fact that in the construction of the focusing system the source of the medium creation in the form of easily ionized substance, which forms the steam-gas medium with low threshold of the breakdown, which is especially important in the case of insufficiently high medium pressure or directly in the vacuum.

17.11 New Approaches for Conductive Channel Creation

The displacement of the optical focusing system in the media and interaction of P–P laser radiation with energy in the focus of optical system sufficient for the breakdown ensures the formation of the continuous current-conducting channel in the air medium due to its ionization, in this case. As calculations show, the frequency band of laser pulses ensures the continuity of the formed current-conducting channel in the appropriate speed range of the moving focusing system, which can be realized in this particular medium. Actually, each pulse of the laser, with the help of focusing system can create a certain extensive region of plasma, which applies to the relatively small section of space in the trajectory of the displacements of that focusing system. If the repetition rate of pulses will follow each other with small frequency the displacements of the plasma regions after focusing system will be something more similar to the dotted line. However, with higher frequency, for example more than 10 kHz, and optimum for the medium speed the displacements of ionization regions will no longer have breaks and the formed by this process conducting channel will be continuous (Fig. 17.5).

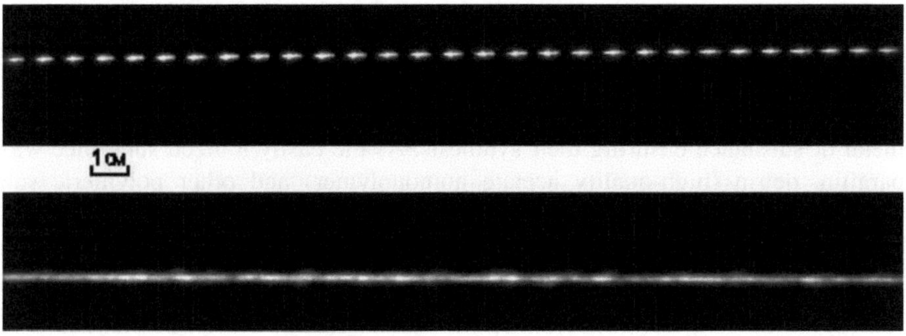

Fig. 17.5 Time controlled evolution of trains in the space for F = 20 Hz, 50 cm/s and for F = 25 kHz, 250 cm/s correspondingly

However, after entering of moving optical system into much less density layers of the atmosphere and further into the vacuum the problem of the scarcity of medium appears, which could complete a deficiency in the ionized gas both as the source of conducting medium and the means of shock wave accumulation and transfer to the object of the motion of necessary quantity of energy. For this case the moving optical system must be supplied with the source of medium creation in the environments of laser focus in the form of sublimated under the action of incident laser radiation easily ionized substance, containing the nano-particles of metal or substance, which under the action of laser emission ensures the synthesis of such a nano-particles. The presence of easily ionized substance source, which due to the sublimation process under the action of laser emission, makes it possible to form the current-conducting channel in the vacuum ensures the continuity of the canal in this case. Presence of nano-particles of metal in the sublimated substance allows simultaneously with improvement of channel conductivity conditions to increase the value of the specific impulse of thrust of laser jet engine (LJE), containing mobile optical system and reservoir with the sublimated substance, which in the essence with this use has determined the time of reaching the required length of electrical power transmission channel [5].

The introduction of high voltage electrode into the current-conducting channel, formed by the moving optical focusing system, allows, from one side, to stabilize the conducting channel of the required length and direction and from the other side to transfer energy by this channel very effectively. In order to facilitate the breakdown conditions of the medium in the optical system focal point with the appearance of not conducting channel only but high energy shock wave as well, it is expedient to use the easily ionized substances, which form the steam-gas medium as the source of medium creation with low breakdown threshold. The latter fact makes it possible to decrease the requirements for the laser pulse breaking energy value.

The moving focusing system can be executed in the form the conical body or combination of cylinder and cone, inverted by apex to the side of direction of its motion. In the end of moving system special optics is established, which can be selected from the number of known optical systems, and ensuring the focusing of the laser light, which falls down to the optical system at the certain distance from the surface. In the body of moving system the sublimated under the action of laser emission easily ionized substance is placed, which contains the nano-particles of metal or substance ensuring their synthesis. As the easily ionized substance wax, paraffin, delrin (high-quality acetate homopolymer) and other polymeric substances can be used. The nano-particles of metal can be of significant dimensions, for example 10–100 nm and can be prepared from the metals such as aluminum, tin, copper, Na, K, Cs, lead and so on. The carbon-containing substances in the combination with the alkali metals can be used as the substance, which makes it possible to synthesize the nano-particles of metal or the metal-composite materials, which possess sufficiently high electrical conductivity. From the literature they are well known—fullerenes on the basis of alkali metals, fullerenes on the base of coppers, which demonstrate high-temperature superconductors properties with tc > of 140 K. For obtaining the nano-crystalline powder the plasma and laser methods of heating can be used. Thus, the nano-particles of carbides, oxides and nitrides with the aid of the pulse laser heating of metals in the rarefied atmosphere of methane (in the case of carbides), oxygen (in the case of oxides), nitrogen or ammonia (in the case of nitrides) were obtained. The pulse laser evaporation of metals in the atmosphere of inert gas (He or Ar) and gas-reagent (O_2, N_2, NH_3, CH_4) makes it possible to obtain the mixtures of the nano-crystalline oxides of different metals, oxcides, nitrides or carbide—nitride mixtures. Composition and the size of nano-particles can be controlled by change in pressure and composition of inert gas atmosphere and gas reagent, by power of laser pulse, temperature gradient during the cooling process. Two reasons for the fine powder to be more conductive can be taken into account. One of them is the strength of electric field increase due to the size and shape of the particles. Another one is the concentration of particles in the dust plasma track of "Impulsar" [2, 3]. The content of nano-particles in the sublimated substance usually composes 10–15 % of the mass. As the source of laser emission high repetition rate P–P laser should be used. The moving focusing system is started directly from the earth's surface with the aid of high repetition rate P–P laser emission with the energy sufficient for the medium breakdown in the focus of optical system (Fig. 17.6).

As a result of laser radiation concentration in the relatively small volume the breakdown of medium takes place. Air breakdown appears in the region of ionization, which applies to a certain distance following the moving focusing system, and the shock wave, which gives to a moving system additional momentum, leading to its acceleration. Upon the transfer of moving focusing system into rarefied layers of the atmosphere or open space, breakdown and appearing in this case shock wave will be achieved in the easily ionized substance, which the moving system has been supplied in advance. With the arrival of the moving focusing system up to the assigned distance or on reaching by the current conducting channel the energy receiver it is necessary to place the electrode of high voltage source into the canal and to ensure the energy transfer process. The conductivity of channel proved to be sufficient for the discharge of capacitor bank to the ground just

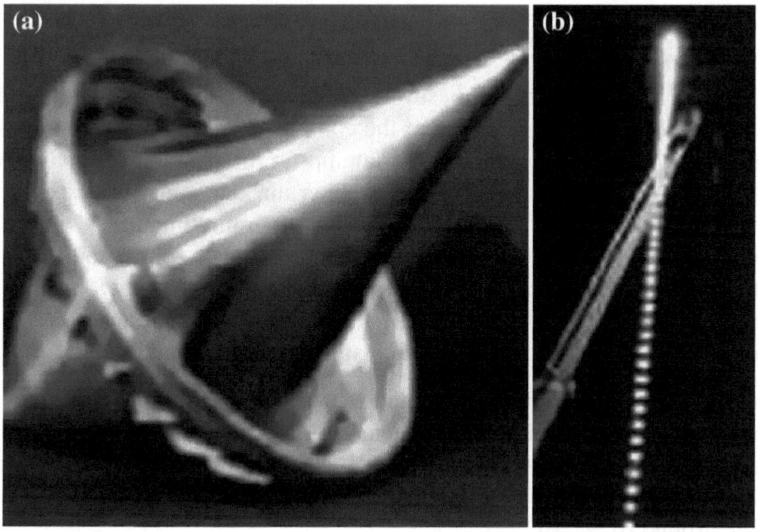

Fig. 17.6 Launch of vehicle by low rate P–P laser (Rep. rate –20 Hz, P = 10 kW)

after reaching by the moving optical set of the grounded electrode, located at the final point of lift.

The results of dust plasma electrical breakdown tests are presented here without detailed explanation of physical matter of experimental arrangements. The most simple and effective dust plasma production methods were involved into our investigation. Best result of specific voltage for dust plasma breakdown was measured for electrical Cu wire explosion approach. The results of dust plasma modeling are [6]:

7. P–P CO_2 laser with train of pulses and total energy—20 J. Dust particles size 50–250 nm and concentration N = 10 to the 10–12 power.
   - For dust plasma produced by laser ablation: paraffin wax + graphite or permalloy powder –300 V/cm,
   - For nano-particles of Al, Fe, segnetoelectrics –350 V/cm.
8. CO_2 single pulse laser channel, produced by conical optics—100 ohm/m, 100 V/cm;
9. Long channel (filaments) produced by solid state laser—0.1 Mohm/m, 7.5 kV/cm;
10. Long channel produced by wire explosion—10 ohm/m, 45–50 V/cm;
11. Data for wire explosion channel, corresponding to the delays: 10 mks—50 mks—75 mks; 97 V/cm—124 V/cm—166 V/cm.

It should be mentioned here that min. resistance value is very much dependable on cross-section of dust particles sheaf, concentration and size of particles, repetition rate of laser pulses, average power of laser, gas pressure, humidity and many other parameters are involved.

17.12 Conclusion

In conclusion it is necessary to say several words about the interrelation of atmospheric electricity with the formation of the composition of atmosphere and climate of planet. Many years were undertaken attempt at the creation of the conducting channels of large length for studying the upper air and solution of special problems. In this connection is of great interest the program "Impulsar", which in the combination with the high-voltage high-frequency Tesla's source can be useful in the solution of the enumerated above problems. At the same time one should say that as a result of the conducted investigations LJE is possible the passage to the solution of the following completely interesting problems:

1. Creation of the interceptor of manmade space debris and other dangerous space objects, such as asteroids, comets, meteorites and so on [7].
2. Launch of micro-objects with the super-high acceleration to the space [8];
3. Realization of orbital scale conducting channels for energy delivery from space to the ground [6, 9, 10].

Powerful pulsed lasers are capable to create a large scale current conducting channels, which can be located at the arbitrary distances from the source of radiation. Channel with the length about hundred of meters can be achieved with low energy of single laser pulse. Beginning from the 70s, successful attempts of the problems: interception of lightning bolts and blocking of over voltage waves on the electric power lines were undertaken.

Successful developments of high repetition rate P–P powerful lasers technology and technology of "Impulsar" system make it possible to foresee the possibility of well conducting channels realization with the length up to several 10 and hundred of kilometers for the purpose of energy transfer for significant distances, creation of new and promising systems for the mastering of outer space power engineering and motivation for significant contribution to be done on that basis to an essential improvements of the global ecology of our planet.

References

1. V.V. Apollonov, Yu.S. Vagin, Conducting channel creation in a non-conductive media. RF Patent № 2009118874, 2009
2. V.V. Apollonov, Yu.S. Vagin, A.B. Egorov, V.V. Kiyko, V.I. Kislov, Patent РФ № 2175159
3. V.V. Apollonov, High power lasers for "Lightcraft" applications. in *Proceedings of ISBEP-6*. Santa Fe, NM, USA, 2009
4. V.V. Apollonov, *Vpered k Tciolkovskomu* (RTSoft, Moscow, 2010)
5. V.V. Apollonov, Yu.S. Vagin, A.B. Egorov, V.V. Kiyko, V.I. Kislov, High-frequency pulse-periodic mode for high-power laser. in *Proceedings of ISBEP-2*, Sendai, Japan, 2003
6. V.V. Apollonov, N.V. Pletnev, Formation of extended conducting channels in atmosphere. QE **42**(2), 130–139 (2012)
7. V.V. Apollonov, *High Power Lasers for Space Debris Elimination, HPLS@A-2012* (Istanbul, Turkey, 2012)

8. H.T. Su, R.R. Hsu, A.B. Chen, Y.C. Wang, W.C. Hsiao, L.C. Lee, M. Sato, H. Fukunishi, Gigantic jets between a thundercloud and the ionosphere. Nature **423**, 974 (2003)
9. V.V. Apollonov, To the space by laser light. Vestnik of RANS **1** (2008)
10. V.V. Apollonov, V.R. Sorochenko, Yu.A. Shakir, Experimental investigation of resonance single wire line of energy transfer. Report of GPI RAS (2009)

Chapter 18
Jet Engine Based Mobile GDL—CO_2 for Water Surface Cleaning

18.1 Introduction

What is the best to start this chapter with? "Deep water Horizon" (BP's operation) case in the Mexican gulf is the best example of man made natural disasters. Today, the notion that "offshore drilling is safe" seems absurd. The Gulf spill harks back to drilling disasters from few decades past—including one off the coast of Santa Barbara, California in 1969 that dumped 3 million gallons into coastal waters and led to the current moratorium. The "Deep water Horizon" disaster is a classic "low probability, high impact event"—the kind we've seen more than our share of recently, including space shuttle disasters, 9/11, Hurricane Katrina and earth quake in Japan. And if there's a single lesson from those disparate catastrophes, it's that pre-disaster assumptions tend to be dramatically off-base, and the worst-case scenarios downplayed or ignored. The Gulf spill is no exception. Fire boats battle the fire on the oil rig "Deep water Horizon" after the April 21 terrible explosion.

The post-mortem is only the beginning, so the precise causes of the initial explosion on the drilling platform and the failure of a "blowout preventer" to deploy on the sea floor probably won't be established for weeks or months. But the outlines of serious systemic problems have already emerged, indicating just how illusory the notion of risk-free drilling really was, while pointing to some possible areas for reform. A "blowout" on an oil rig occurs when some combination of pressurized natural gas, oil, mud, and water escapes from a well, shoots up the drill pipe to the surface, expands and ignites. Wells are equipped with structures called blowout preventers that sit on the wellhead and are supposed to shut off that flow and tamp the well. "Deep water Horizon's" blowout preventer failed. Two switches—one manual and an automatic backup—failed to start it. When such catastrophic mechanical failures happen, they're almost always traced to flaws in the broader system: the workers on the platform, the corporate hierarchies they work for, and the government bureaucracies that oversee what they do. According to the study 600 major equipment failures in offshore drilling structures 80 % were due to "human and organizational factors," and 50 % of those due to flaws in the engineering design of equipment or processes. With near-shore

V.V. Apollonov, *High-Power Optics*, Springer Series in Optical Sciences 192,
DOI 10.1007/978-3-319-10753-0_18

and shallow reserves of fossil fuels largely depleted, drilling has moved farther off shore, into deeper waters and deeper underground. The technology for locating oil and gas reserves and for drilling has improved, but the conditions are extreme and the challenges more formidable. This is a pretty frigging complex system. You've got equipment and steel strung out over a long piece of geography starting at surface and terminating at 18,000 ft below the sea floor. So it has many potential weak points. Just as Katrina's storm surge found weaknesses in those piles of dirt—the levees—gas likes to find weakness in anything we connect to that source.

Why we have paid in our paper such a detailed attention to the policy of BP's "Deep water Horizon"? The contamination of large water areas (oceans, seas, lakes, and rivers) with petroleum products as a result of accidents and not so smart industrial activity of similar companies and officials around them is one of major problems of protection of environment. Any other contaminant can not be compared with petroleum on a basis of universal utilization, number of contamination sources and degree of effect on all components of environment. Through penetration of petroleum products into water, there are deep, frequently irreversible changes of its chemical, physical and microbiological and even global (redirection of the ocean streams) properties.

The following methods of disposal of petroleum contamination of water: natural, mechanical, physical-chemical, chemical, biochemicals are widely used nowadays.

The laser method of cleaning of water surface from a thin petroleum film is one of physical-chemical methods. Some time later after penetration into water, the petroleum spreads on a surface of a water and forms very thin film (thickness several microns). This film cannot be forced to burn, since because of a good thermal contact to a surface of water the film cannot be heated to temperature ensuring steady combustion.

The principle idea of a laser method consists in following. The laser beam passes through a petroleum film and then is absorbed in thin layer of water. The water heats very fast up to the boiling temperature, and the forming vapor destroys a film, bursting it in small-sized fragments, which, mixing up with the hot vapor, are decomposed quickly with formation of simple un-toxic substances.

The main advantages of a laser method consist of the following.

(a) This method is "fast-response" one, since does not require any special preparation; in emergency situations the time from the moment of obtaining of the alarm signal to the beginning the laser installation operation is determined only by time necessary for arrival of the ship or the helicopter with the installation on board in given area.
(b) The method is contactless, i.e. does not require realization of preparatory or other activities in the oil spillage.

For realization of the proposed technology, it is possible to utilize different types of lasers, as continuous (power up to 250 kW), and high repetition rate pulse-periodical (average power of the same level with duration of pulses 100 ns and repetition rate up to 100 kHz).

Theoretical estimations and the experiments have shown, that as a result of a laser beam action on a surface of a water and land, covered with film of hydrocarbon contamination, following effects can be observed:

Evaporation and burning, and in a continuous mode the consumption of energy for 1 gm of vaporized liquid on the order of value surpasses energy necessary for heating up to boiling temperature and evaporation of a film, that is explained by a heat consumption for heating of water. In a pulse mode, the consumption of energy it is per unit of vaporized mass of a liquid (film) approximately in 5 times less, thus the process of film ignition start more easily.

Knocking out the particles of polluting substance above a surface of water under action of pulse or powerful scanning laser radiation should be considered [1]. The physical mechanism of this effect is explained by sharp evaporation (boiling-up) of a thin layer of water under a layer of polluting substance. This process takes place where as hydrocarbon polluting thin film has an absorption coefficient less, than water. Knocked out particles of petroleum at a height up to 50 cm is possible. For this case, the energy consumption per unit mass of a raised liquid is sufficiently less than energy necessary for evaporation and ignition (mode number one) or for sucking up and saving for future efficient usage (mode number two).

The character of a task to be solved superimposes certain conditions on operating characteristics of the laser device concerning both parameters of laser source, and concerning auxiliary systems.

Technological lasers of 10–15 kW power range, that widely are used in production, will have the output completely insufficient for liquidation of large scale contamination presenting the greatest danger according to the above-stated estimations. Such lasers can be utilized, at the best, for improvement of a process engineering of cleaning under modeling conditions.

Besides, the character of this task dictates impossibility or extreme undesirability of use of the stationary civil engineering service line (electric power network, water pipe, main gas line etc.), in view of the requirement of a high self-sufficiency and mobility. The power unit should not limit mobility and thus provide totally energy needed of all installation, the necessary reserve of expendables and fuel on board a complex. A capability of a fast redeployment from one type of a vehicle on another is also desirable.

In the present technical paper, the basic design concept of the mobile laser installation on basis of gas dynamic CO_2-laser is developed.

18.2 The Substantiation of Selection of Laser Type According to Specific Parameters, Operation Autonomy and Mobility

There are few the most effective and scalable modern high power continuous/pulse-periodical operated gas lasers should be considered for realization of clamed in the title of that paper tasks: electro discharge laser (EDL), gas dynamic laser (GDL), chemical laser HF/DF (CL) and chemical oxygen—iodine laser (COIL). We are not going here to details of high power high repetition rate P–P laser systems operation. Main results of oil films elimination detailed consideration taken in the past can be summarized and reduced to the paper format and presented here as following:

18.2.1 Analysis of Applicability of Various Types of Lasers to the Task

The laser with output power up to 250 kW of continuous operation (CW) or high repetition rate pulse-periodical operation (P–P) during few hours (minimum requirement) is required for realization of the mentioned above task. Besides of that, according to the physics of the process of destruction of a petroleum film, explained in Introduction, laser radiation should be weakly absorbed by petroleum film and should be absorbed effectively by water. In the Table 18.1, the computational data for depth of radiation penetration for four types of mentioned above lasers in petroleum and in water are presented.

The range of values of depth of penetration for petroleum is the sequence of the fact, that absorption coefficient of various grades of petroleum and water (for the CCL case) in relation to wavelength in CCL radiation spectral band were considered.

From the Table 18.1, it is obvious, that the COIL radiation is rather weakly absorbed by water and is rather strongly absorbed by petroleum. However to heat a petroleum film by thickness ~100 μ through absorption of COIL radiation practically is impossible because of strong heat transfer from a film into water through heat conductivity. The conclusion is, to use COIL for the solution of this task is impossible.

As the lengths of waves of radiation GDL and EDL are identical, the comparison of these two types of lasers is carried out on the basis of mass—dimension characteristics and other parameters that are listed in Table 18.2. The comparison is carried out for two real installations, which were developed and tested in the past to sufficient degree, precisely—GDL of rated power 100, and 20 kW power EDL.

The analysis of the data, presented in Table 18.2 shows, that GDL has decisive advantages in comparison with EDL. It is necessary to pay attention, that EDL specific fuel consumption is three times less, than that of GDL. However influence of this factor on the total complex weight will have negative influence only for large operation time (5 h or more). At the arrangement of a laser complex on the board of helicopter, the operating time will not exceed 0.5–1 h, therefore in this case smaller EDL specific fuel consumption will not have decisive value.

Table 18.1 The computational data for depth of radiation penetration for 4 types of lasers

The laser type	Depth of penetration (μ)	
	Petroleum	Water
GDL, EDL	100–260	10
CCL	26–200	0.8–40
COIL	50	2000

Table 18.2 Comparison of basic parameters for two lasers

No.	The name	GDL	EDL
1.	Power laser (max) (kW)	100	20
2.	Overall dimensions of the laser installation (m)	9 × 2.4 × 2.4	2.5 × 3 × 3.5 + 2 × 0.8 × 0.8
3.	"Dry weight" (free of fuel) (metric ton)	6	18.6
4.	Specific, 1 kW of laser power dry weight (t/kW)	0.06	0.93
5.	Specific, 1 kW of laser power, dry volume (cubic m/kW)	0.5	1.5
6	Specific volume fuel consumption for 1 kW of laser power (cubical m/h/kW)	0.03	0.01

Thus, GDL and CCL remain only as competing systems. The comparative analysis of these two types of lasers with reference to a considered task is stated below.

18.2.1.1 Specific Power and Wavelength of Laser Radiation

GDL specific power, Wsp is 10 J/g; this value for CCL is much higher and reaches 150 J/g. From this point of view, CCL has conclusive advantage in comparison with GDL.

GDL radiation wavelength, $\lambda = 10.6\ \mu$ and CCL wavelength, $\lambda = 2.7\text{–}3.3\ \mu$. From this point of view, CCL has one more potential advantage over GDL, as the diffraction limited angle of divergence of CCL beam is 3–4 times less in comparison with GDL (at identical aperture). However this problem requires the further arguing.

(a) Radiation extraction from the resonator.

In high-power lasers, extraction of laser beam from the resonator, where the pressure is much lower than atmospheric pressure, in external space through a rigid window is practically impossible for the reason of thermal destruction such window. For a extraction of such beam out of installation, the aerodynamic window (AW) usually is utilized. The supersonic gas stream cross-sectional to beam fulfills the task of transparent boundary. As there are areas of various density gas sheaths limited by curvilinear surfaces in such flow, the beam transmission through this gas non-uniformity introduces the distortions of a wave front resulting in increase of an angular divergence of a beam. These distortions do not depend on a wavelength, as the dispersion of an index of refraction of gas is extremely small and therefore these distortions have the same absolute value for GDL and CCL. However relative values of these distortions of a wave front set (i.e., reduced to wavelength) are for CCL 3–4 times higher, than for GDL, therefore from the point of view of an angular divergence of radiation the advantage CCL in comparison with GDL can be not so essential.

(b) Use of an external optical system.

The optical systems GDL and CCL for delivery beam energy to remote objects include necessarily external (in relation to the resonator) optical elements and systems, intended for expansion of a beam with the purpose of reduction of its divergence. A principal component of a telescope is the main mirror, which diameter can reach several meters (depending on required power density on the target). The one piece glass or metal mirror will be too heavy. For this reason the main mirror is usually considered as consisting from large number of small mirrors (facets), assembled in a unit. Every facet has the own adjusting device. The accuracy of relative positioning these facets should be not worse $\lambda/4$, otherwise distortions of a wave front after reflection of a laser beam from this mirror will cause the essential increase of an angular divergence in comparison with it diffraction limit. This is $\lambda/4 \cong 2.5\ \mu$ for GDL, $\lambda/4 \cong 0.75\ \mu$ for CCL. Therefore in practice required accuracy is much more difficult to ensure for CCL.

This brief discussion shows, that the efficiency of CCL external optical system can be much below, than that of GDL and, hence, reduces advantage of shorter wavelength of CCL radiation.

Besides at small length of a beam paths from the laser down to a water surface, the angular divergence has no importance, since density of power on a water surface appears sufficiently high without application of an external optical system.

18.2.1.2 Length of a Zone of Radiation Generation

In GDL working section (the resonator region) the transfer of oscillation energy from N_2 to CO_2 has taken place. The characteristic time of this process is relatively large. Besides the relaxation time of oscillatory—exited molecules CO_2 at their collisions with other particles is also large. Therefore for full extraction of N_2 accumulated oscillatory energy, and transformation it in laser radiation, length of a working zone in flow down direction should be 10–15 cm. Such large length provides certain freedom of selection of the resonator scheme (one pass or multipass, symmetrical or asymmetrical, etc.). Thus the mirror operating surfaces are large enough, that results in a reduction of its thermal loading and distortion of a surface due to inhomogeneous heating.

In CCL, the length of beam generation section is small, less than 2 cm. This circumstance causes difficulties in optimization of the laser resonator and leads to increase of thermal loading of mirrors, that results to significant thermal distortion of mirrors and distortion of a wave front, i.e., in increase of an angular divergence of radiation.

18.2.1.3 Direct Exhaust of Used Gas to Atmosphere

If GDL or CCL is placed on ground or on board of flight vehicle (plane or helicopter), there is a problem of exhaust of spent gas in atmosphere. In a case GDL this problem is decided rather simply by use of the diffuser, as the operating pressure in a zone of the resonator is rather high. For reaching a maximum degree of pressure recovery in the diffuser it should have special geometry.

In CCL case the operating pressure in the resonator is much lower in comparison with GDL, therefore the direct exhaust in atmosphere with the help of the diffuser is impossible, it is necessary to utilize additionally gas or water vapor ejector to ensure necessary pressure in output section of the diffuser. Such ejector requires large additional gas flow rate, that results in fundamental complexity of the laser installation design and significantly decrease specific power of generation, i.e., in this case one of main CCL advantages is lost in comparison with GDL.

18.2.1.4 Toxic Characteristics

The degree of danger CCL from the point of view of toxic characteristics of working components and exhaust gases is much higher in comparison with GDL. In CCL, for creation of a working mixture the extremely toxic substances containing fluorine will be utilized. Besides the exhaust CCL contains a significant quantity of fluorine hydride HF, which is supertoxicant. In this connection with use CCL in a system of laser cleaning of water areas it will be necessary to ensure special safety measures, that, naturally, will cause the complication of the installation design and rise of its price.

In a case GDL exhaust gas is ecological clean ($N_2 + CO_2 + H_2O$). However, CO, carbon monoxide which is toxic gas can be used as a fuel for GDL. Hence, system of storage and supply of CO should ensure absence of its leakage. On the other hand, CO is lighter-than-air, therefore it floats in atmosphere and is fast blown away by wind.

If as fuel in GDL will be used liquid hydrocarbon fuel (benzole, toluene), the system of its storage and supply does not require special measures to secure the ecological safety.

18.2.1.5 Absorption of Laser Radiation in Atmosphere

The wavelengths of GDL and CCL radiation are within of windows of atmosphere transparency. These windows are usually determined for vertical transmission of beam through atmosphere, and for this case power dissipation on track is ~20 % for GDL and 25 % for CCL. For vertical tracks of length 50–150 m, which are of interest with reference to a considered problem, the radiation power dissipation can be neglected.

18.2.2 Substantiation of Selection Gas Dynamic CO_2-Laser

Proceeding from the comparison of various type lasers, explained in the previous chapter, it is possible to make a conclusion, that the most suitable laser for development of the installation for disposal of petroleum films is GDL for following reasons:

18.2.2.1 Simplicity of a Design

From design point of view GDL components are most simple of the considered lasers. The combustion chamber operates on usual components, for example, kerosene + air. The designs of such chambers are practically perfect, there is a wide experience of their operation in various technical devices. The nozzle unit is GDL made from heat resisting steel and the long time without forced cooling can work. The long zone of population inversion behind the nozzle unit (10–15 cm) stipulates simplicity of selection of the scheme of the resonator for obtaining laser generation with maximum efficiency. Thus the characteristic size of a mirror of the resonator is approximately equal to length of a zone of generation, i.e., 10–15 cm, therefore manufacturing of these mirrors does not call serious technological difficulties.

The capability of GDL activity at high pressure of gas in the resonator in front of the nozzle unit (up to 3 MPa and more) provides with a straight line an exhaust of spent gas in atmosphere with the help of the supersonic diffuser of a special design. Such capability all remaining considered lasers are deprived. The aft ejector is necessary for maintenance of a direct exhaust of these lasers which requires additional gas flow rate, that complicates the installation and reduces specific energy of radiation.

It is necessary to note also, that GDL output power rather weak varies at change over a wide range of parameters of gas (pressure, temperature, chemistry) in the combustion chamber, i.e., GDL is not critical to accuracy of a task and maintenance of an operational mode.

One more virtue GDL is the small time (some seconds), necessary for start.

Thus, GDL is the rather simple, reliable and flexible tool ensuring high output power of laser radiation.

18.2.2.2 Mobility

From all considered types of lasers GDL for today has, apparently, least "weight" and "volume" of 1 kW of laser power. It allows rather simple to place GDL on any vehicle, whether it will be airplane or helicopter, ship or railway platform, etc. Such GDL mobility allows with its help to solve many tasks, including connected with disposal of petroleum films.

18.2.2.3 Selection of Propellant Components

One of specific singularities of working process of the CO_2-laser is, that the working mixture of gases should contain in a receiver rather small quantity of water vapor (no more than 5–6 volume %). On the other hand, the operation temperature of gas in a receiver should be enough high (1,300–1,600 K). It superimposes limitations on selection of propellant components (fuel and oxidizer). From the point of view of maximum specific energy of radiation (J/kg) optimum components are gaseous at standard conditions with damp (CO) and liquid nitrous oxide

(N_2O), thus additional nitrogen (or air), necessary for working process, moves in a receiver from a separate source, and the water vapor in quantity 1.5–2 % is formed by burning small quantity of hydrogen or alcohol in air.

18.2.2.4 Wavelength of Radiation

The analysis of the data presented in Table 18.1 show, that for the solution of a problem of disposal of petroleum films, the optimal wavelength of radiation is that of the CO_2-laser, as this radiation is weakly absorbed by the film and is strongly absorbed by water. From the point of view of a diffraction limited angular divergence of radiation, the CO_2-lasers (EDL and GDL) lose for CIL and CCL. However with reference to a considered problem, the value of a divergence is not critically important. The estimations show, that at length of a beam from the laser source down to a surface of water ~50–100 m and power of radiation ~100 kW, an precision focussing of a beam is not required, for the reason of obtaining required density of power in a spot on water surface. It is possible to utilize nonfocussed or partially focussed beam. Therefore with reference to a considered problem, the relatively large wavelength of GDL radiation is not an insufficiency factor.

The explained above reasons allow to make a unequivocal conclusion for the selection GDL as the generator of laser radiation for the mobile installation intended for disposal of a petroleum film on a surface of water. At the estimated power of radiation, GDL is preferable in weight and volume factors in comparison with EDL, HCL and COIL: it does not require the electric power, will utilize low toxic fuels, provides a direct exhaust of utilized gas in atmosphere. GDL simple in control, it is not critical to changes of working parameters in sufficiently broad range, and it is convenient in operation. The important circumstance is that it exists reliably operable GDL prototype with power range of 100 kW, on which the extensive researches are carried out. In particular, the design of an aerodynamic window permitting to extract a laser beam from a zone of the resonator in atmosphere without application rigid transparent for a working wavelength of materials was developed.

18.3 Selection and Substantiation of Basic Performances of the Laser Installation and GDL Principal Diagram

In the previous chapter the comparison of various types of high power lasers was carried out from the point of view of their application for the solution of this task, where was shown, that the most appropriate type of the laser is CO_2—GDL. However there are different types of GDL, therefore it is necessary to choose among them optimal version, and also to develop the general concept of the power installation for optimal supply of working components in GDL.

In the present chapter the analysis of the various GDL schemes, their advantages and disadvantages are given from the point of view of application in the considered installation, the selection of propellant components for GDL is justified from the point of view of their power efficiency, production and toxic characteristics, and ecological safety.

Last portion of the chapter is devoted to the substantiation of the GDL characteristics working on the chosen components. For this purpose the calculation results which have been carried out according to the report [2], permitting to define specific power of laser radiation at the given initial parameters, such as fuel chemistry, temperature and pressure in front of the nozzle block, expansion ratio of the nozzle etc. are used. The calculation results and their analysis are shown.

For definition of the main dimensional characteristics of the laser and its flow organizing parts it is necessary to define a basic level of power of laser radiation, which, in turn determines the total fuel consumption and remaining geometrical characteristics of the laser.

18.3.1 Selection of Optimal Radiation Power of GDL

18.3.1.1 Theoretical Substantiation of Power Level

The basic capability of disposal of petroleum contaminations from water surface by means of irradiation of a petroleum film by a high flux of laser radiation is proved experimentally in our publications. However, the published data give only the general representation about basic possibility of using of such method. It is obvious, that the combustion is complex multiparameter process and, if we talk about operational use of the mentioned principles in the particular technical device intended for practical use, the optimization of parameters of the installation is necessary. The important value in this case is the efficiency, which it is possible to understand as the area of cleaned water surface divided to mean energy, used on clearing, (m_2/kW). For optimization of parameters of the installation it is necessary to conduct a cycle of research works on study of evaporation, ignition and combustion processes of petroleum used for the above mentioned method. The experimental research of the given problem even without qualitative understanding of involving processes represents a very difficult task in connection with vast quantity of optimization parameters and requires heavy material costs and time. In this connection the preliminary development of theoretical model representing process of disposal of petroleum film is desirable.

In [3] presented for the second phase, the various interaction mechanisms of a laser beam with a petroleum film on a water surface are theoretically considered. The following mechanisms were included:

- laser heating active absorbing, translucent and transparent films;
- an explosive boiling-up and conditions of process efficiency;
- evaporation and vaporization.

The basic conclusions of the technological report are the following.

1. For reaching temperatures of evaporation of a film the necessary time of radiation action is about one milliseconds at laser power 50 kWE and diameter of an irradiated zone 10 cm. Temperature of water layer under the surface of film can essentially exceed boiling point. Such situation arises due to superior velocities of laser heating (up to several millions degrees of Celcius per one second) and inertial character of heat transfer process. At such temperatures the water passes in a metastable state with active outgassing. Under a surface of a film there will be microexplosions of air—steam bubbles. As result, there will be a separation of a film from a surface of a water, its breaking and outflow of drops of petroleum in air.
2. The scanning velocity at power of the laser of 50 kW and diameter of a spot 10 cm ought to be in range of 10–12 m/s. Thus the velocity of cleaning of a surface 6,000–8,000 m^2/h is provided.

It is necessary to note, that all calculations are fulfilled for thickness of a petroleum film 100 μ, that is rather large value and is observed only in the initial moments of formation of oil spillages. At thinner films, the velocity of cleaning will increase proportionally.

18.3.1.2 Experimental Researches

For experimental confirmation of basic physical principles included in the basis of the Project, the participants carried out preliminary experimental research of interaction of laser radiation with films of various petroleum types on a surface of water. The research works were conducted within the framework of financing of activities.

During research works gas dynamic laser developed by JSC "Energomashtechnika", and established on the bench was used.

The optical scheme of realization of experiments is presented in Fig. 18.1.

The gas dynamic laser 1 generates radiation 2 with a wavelength 10.6 μ. The radiation is directed on a concave focusing mirror 3 and, being reflected, impacts on a surface of test heterogeneous structure in a cuveta 4. The mirror realizes periodic oscillations with the help of the mechanical device 5, thus the laser beam scans on a surface of a dish.

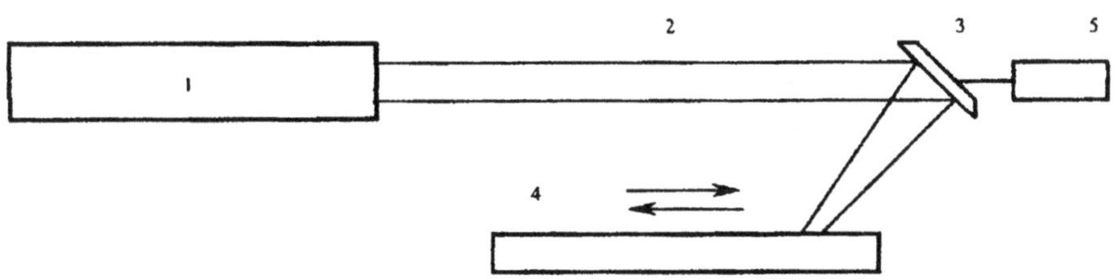

Fig. 18.1 Scheme of experiment

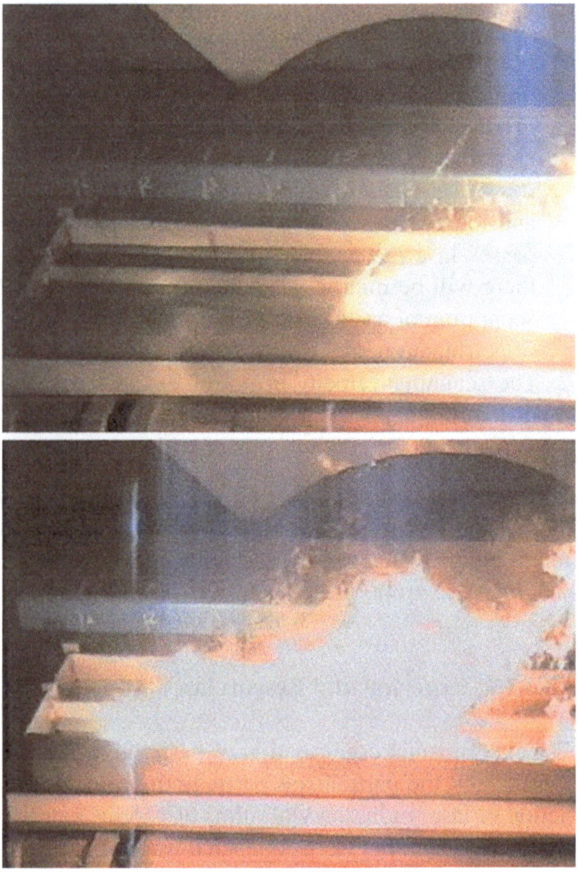

Fig. 18.2 Intensive combustion of petroleum contamination film under laser action

Diameter of laser beam cross section on a surface of a dish changed from 5 up to 10 cm. The laser radiation power was within the limits of 50–100 kW.

In Figs. 18.2, 18.3 and 18.4 the results of experimental research works are presented, on which it is possible to make following preliminary conclusions.

1. The carried out experiments testify the possibility of using of the laser installations for cleaning of an aqueous medium from contaminations of petroleum and of it processing products.
2. For definition of optimal parameters of the laser installation and operation modes in practical conditions it is necessary to continue experimental research works under the special program approved by a potential customer.
3. For necessary reserve, the radiation power of gas dynamic laser of 100 kW for the installation to be developed according to present technical proposal is chosen.

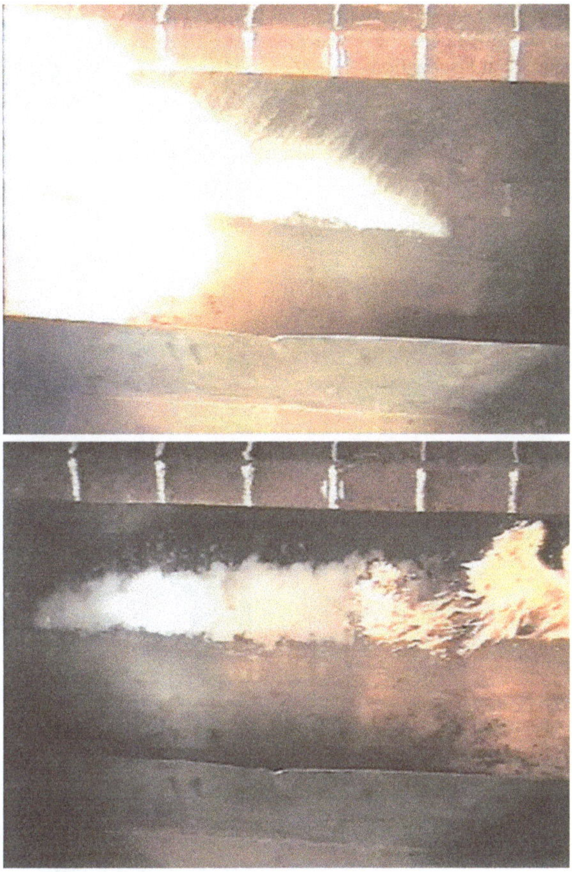

Fig. 18.3 Combustion of a film of kerosene at laser action

18.3.2 Selection of the GDL Scheme and Its Substantiation

18.3.2.1 Conventional Scheme GDL

The principle of GDL operation is based on non-equilibrium fast expansion of the heated up working gas in the supersonic nozzle, at which there is a partial freezing population of oscillatory levels of molecules CO_2 and N_2 and formation of population inversion of upper and low laser levels. The heating of gas to an operation temperature 1,300–1,700 K at pressure 2–3 MPa is conducted in the combustion chamber through of combustion of chosen fuel—oxidizer components. The transformation of the accumulated oscillatory energy of gas to laser radiation has taken place in the optical resonator located down the nozzle. After the resonator the working gas passes through the supersonic diffuser and will be exhausted in atmosphere.

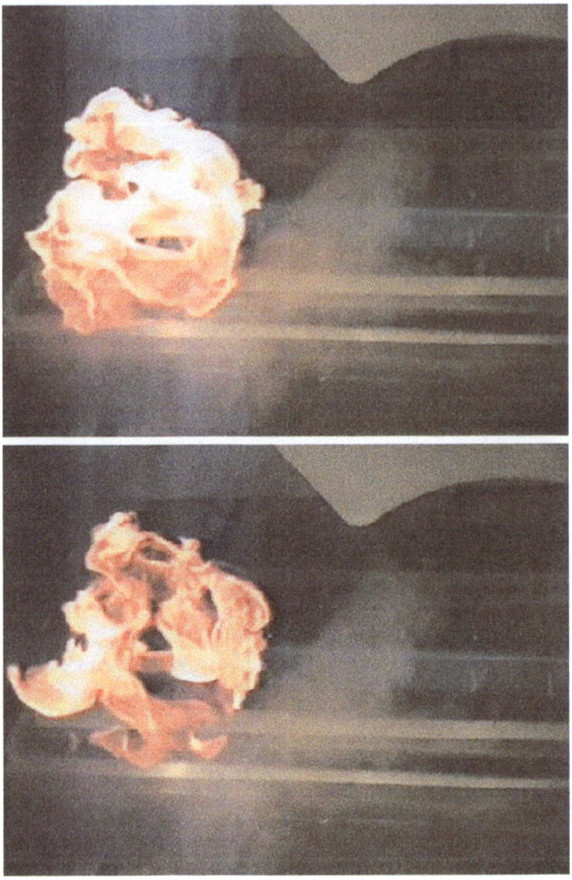

Fig. 18.4 Fire of used engine oil vapors above water surface at laser action

The most fast gas expansion is near to critical cross-section of the nozzle on length equal on an order of magnitude to the size of critical cross-section. Therefore for fast freezing of oscillatory energy it is necessary to use the nozzles with a small size of critical cross-section (<1 mm). On the other hand, the pressure of gas after the nozzle (i.e., in the resonator region) should be small enough (<100 mm Hg) to reduce to a minimum the losses of oscillatory energy due to collision relaxation. It is solved with use of nozzles with large expansion ratio (area ratio of output to critical cross-sections $A/A^* \geq 20$, that corresponds to value of a Mach number M 4).

For effective transformation of the accumulated oscillatory energy to laser radiation the amplification of a small signal during the pass the length of an active zone should be large enough to compensate optical losses (removal of radiation, absorption in mirrors, dissipation in medium) and to ensure a high enough radiation power in the resonator. In GDL of very low powers (tens W) is possible to use the so-called mononozzle, which represents the flat (rectangular) nozzle with critical cross-section as a narrow slot. However for high-power GDL implementation of the mononozzle is practically impossible because of its large length (1 m and more); arising thermal deformations result in essential to nonuniform width of critical cross-section, h*, down to full blocking of critical cross-section in some zones. Besides the mononozzle does not allow to receive enough large transversal size of an active zone, that causes certain problems at resonator selection. Therefore in high power GDL the so-called nozzle unit representing set of a many relatively short (~100 mm) slot-type nozzles is used. These nozzles are formed by contoured nozzle blades made from thermal resisting metal. Such design of the nozzle unit allows to vary over a wide range width and height of an active zone. Besides the nozzle unit provides a uniform enough density field of gas in an active zone, that allows to assure a small angular divergence of GDL radiation.

In view of such advantages of nozzle blocks, they are now widely used in GDL, and therefore it is possible to name such GDL "GDL of traditional scheme". The radiation power of such GDL is up to 100 kW and more in a continuous operation radiation mode obtained.

18.3.2.2 GDL with a Set of Axisymmetrical Nozzles

The nozzle unit of the conventional scheme has the disadvantage that it is complex in manufacturing, and also is subject to influence of thermo-deformations. There is an alternative approach to development of nozzle units, which consists in following.

Instead of two-dimensional nozzles the axisymmetrical nozzles with small diameter of critical cross-section are used. The supersonic part of such nozzle can be made conical or contoured. In this case nozzle unit looks like a plate, in which the numerous small-sized nozzles located close as possible to each other are made. The manufacturing of such nozzles is possible, they are not subject thermo-deformations, by the appropriate arrangement of these nozzles on a plate (quantity of rows and quantity of nozzles in a row) it is possible to give output cross-section of the nozzle unit the required configuration.

However such nozzle unit has at least two disadvantages caused by presence of "empty" zones between nozzles (since complete filling of output cross-section by round nozzles basically is impossible).

The first disadvantage consists that the empty zones are sources of shock waves in a supersonic flow. These waves result in density inhomogeneous structure of gas in a zone of the resonator and, hence, to increase of an angular divergence of laser radiation.

The second disadvantage consists that in empty zones the viscous vortices flows are formed, that results in reduction of total pressure of a main flow. In turn, the reduction of total pressure has negative consequences on the supersonic diffuser operation, that can be appreciated as reduction a pressure recovery coefficient. In this case for realization of direct exhaust of working gas in atmosphere it is necessary to raise combustion-chamber pressure, that results in a decrease of output power of laser radiation.

For these two principles GDL with the nozzle unit as a set of axisymmetrical nozzles is not used at noticeable production scale.

18.3.2.3 GDL with Working Gases Mixing in a Supersonic Flow

In usual GDL a simultaneous heating of all components of a working mixture—N_2, CO_2, H_2O has taken place in the combustion chamber. However initial reserve of laser energy contains, primarily in oscillatory—exited molecules N_2, which part in gas mixture makes ~90 %. The role CO_2 and H_2O consists only of transforming accumulated in N_2 oscillatory energy to laser radiation.

The pure nitrogen has extremely large time of an oscillatory relaxation stipulated by collisions of molecules N_2 among themselves. As the result, the oscillatory levels N_2 are easily frozen at fast expansion of the working mixture in the nozzle. However other components of a mixture (CO_2 and H_2O) considerably accelerate a collision relaxation N_2 and by this reduce a reserve of oscillatory energy.

For increase of a reserve of oscillatory energy it is desirable to raise temperature in the combustion chamber. However in usual GDL temperature is limited to value about 2,300 K; the dissociation CO_2 begins at higher temperatures, and the efficiency of conversion of the accumulated oscillatory energy in laser radiation decreases. In contrast to CO_2 the molecular dissociation N_2 begins at temperature ~4,000 K.

The presented ideas have resulted in the new GDL scheme, which concept consists in following: the pure nitrogen, heated up in a prechamber, expandes in the supersonic nozzle, and at the exit of the nozzle to it are mixed cold CO_2 and H_2O. Such scheme has three basic advantages for the usual scheme GDL: (a) temperature of nitrogen in a prechamber can be essentially increased, (b) more effective freezing of oscillatory energy of nitrogen is reached, (c) energy for CO_2 and H_2O heating is not necessary.

In view of the indicated advantages mixing GDL can basically ensure increase of specific power of radiation many times in comparison with the GDL conventional scheme.

A main problem arising at realization mixing GDL—necessity to ensure fast intermixing hot N_2 with cold CO_2, but the disturbances of a supersonic flow N_2 should be kept minimum.

Some solutions of this problem are possible. Technically most simply to organize mixing of CO_2 and N_2 in wakes. In this case flow CO_2 moves in parallel to flow N_2, in this case two sets of supersonic nozzles—one set for N_2, second for CO_2 are used. A advantage of such scheme of supply is that arising at interaction of wakes disturbances are minimum in comparison with other schemes of intermixing.

The basic disadvantage of this scheme is that the mixing of wakes is rather slowly, therefore for effective work of the laser the working section (from nozzle exit down to the diffuser inlet) should have sufficient length, that results in large relaxation losses and reduction of a pressure recovery coefficient of the diffuser. For these reasons the scheme GDL with intermixing of supersonic wakes, apparently, is not effective.

The theoretical analysis and the experimental data show, that the GDL schemes are more effective, in which the injection of cold CO_2 is made near to critical cross-section of the nozzle in it transonic part. In experiments with heating N_2 in electric arc plasma generating device and supply of CO_2, the data were received record for GDL value of a gain equal 3 % cm^{-1}, that exceeds maximum value of a gain for usual GDL almost to an order of magnitude. However specific power of radiation was insignificant, that is connected to a small scale of the installation.

Despite of basic promising character of mixing GDL, there is a lot of problems connected with their practical realization.

One of the most important problems is obtaining hot nitrogen at temperature 3,000–4,000 K. Application of electric arc plasma generating device for high-power mixing GDL is unreal, since at the consumption of nitrogen 10 kg/s and more required electrical power exceeds 100 MW. It is possible to use for obtaining hot nitrogen special fuels, however all of them are strong toxicants.

As a source of hot nitrogen, the devices can also be considered, in which a combustion of metal in air has taken place. As a result of this strong exothermal reaction hot nitrogen and oxide of metal as particles are formed, which are extracted from the formed flow with the help of a cyclone separator. However necessity of a very high degree of cleaning of a nitrogen stream from particles makes this way problematic.

Other important technical problem is the development of the nozzle (selection of a material, development of a robust design, which would maintain without destruction large heat flows in the region of critical cross-section.

The supersonic diffuser causes one more important problem. For an effective work of the laser translation temperature of working gas in an active zone should be sufficiently low, whereas temperature in a prechamber should be relatively high. For the coordination of these two conditions it is necessary to use supersonic nozzle with large Mach numbers ($M = 5$ and more). However is known, the higher Mach number is, the lower is pressure recovery coefficient for the diffuser. Therefore for a direct exhaust of gas in atmosphere it is necessary to develop special diffusers, thus can appear, that such task basically is impracticable without use of a fodder ejector. However consumption of inducing gas in some times exceeds the consumption of induced gas, and in this case specific laser power designed on summarized gas flow rate, considerably decreases.

From explained above it is possible to make a unequivocal conclusion: taking into account problems connected to development of the high power mixing laser, and also absence of the prototypes of such lasers, for the solution of a task of disposal of petroleum film on water surface it is necessary to choose GDL of conventional scheme.

18.3.3 Selection of Fuel for Mobile GDL

In the present part the analysis of possibility of using various fuels for support of operation of the mobile gas dynamic CO_2-laser (CO_2—GDL), intended for disposal of petroleum film on a surface of water is carried out.

CO_2—GDL provides obtaining high power laser radiation generation in a continuous mode with a wavelength of radiation 10.6 μ. A basic physical principle of operation gas dynamic of the CO_2-laser is the fast expansion in the supersonic nozzle of a mixture of gases (CO_2, N_2, H_2O) preheated up to temperature

1,500–2,000 K. In a supersonic flow with Mach number M = 4–5 temperature of gases down to values 300–350 K, necessary for obtaining, population inversion of laser levels of a molecule CO_2. For nozzle operation at direct exhaust in atmosphere after pressure recovery in the diffuser (the static pressure in a zone of generation is equal 5×10^3 to 1×10^4 Pa) the nozzle inlet pressure (stagnation pressure of a flow), equal 2–3 MPa is necessary. At small transversal and longitudinal sizes of the supersonic nozzle (the height of critical cross-section is equal 0.3–0.5 mm, and length of the nozzle 20–40 mm) characteristic time of the gas flowing through the nozzle appears small and comparable with time of oscillating relaxation of molecules of a mixture of gases, i.e., oscillating—oscillating exchange (*V–V* process) and oscillating—translation relaxation (*V–T* process) has taken place in nonequilibrium mode. As a result of nonequilibrium processes in oscillating exited molecules at certain gas mixture ratio and flow parameters in the supersonic nozzle the inversion population of the upper laser level of molecules CO_2 (0001) is formed. The molecules CO_2, are light generating molecules (laser radiation is realized on transition 0001–1000), N_2 molecules are donors molecules transmitting oscillating energy to the upper laser level of molecules CO_2 (0001), the molecules H_2O are intended for population of the lower laser level of molecules CO_2, (1000). The optimal ratio of mixture of gases for an effective work of gas dynamic CO_2-laser should contain $CO_2 = 0.1$, $N_2 = 0.89$, $H_2O = 0.01$ (volumetric fraction). It is necessary to note, that the increase of a volumetric vapor fraction of (H_2O) results in decrease of the laser radiation power with other things being equal.

For mobile GDL, laser radiation, ensuring high power in a continuous mode, a basic method of thermal excitation is the combustion of such fuel components, which as final products give necessary components of a laser mixture at relatively high temperature. The optimal mixture ratio of gases at high temperature of mixture can be obtained, for example, at burning of carbon oxide (CO) and hydrogen (H_2) in air or other oxidizer (N_2O, N_2O_4) with subsequent use of ballast nitrogen.

The specific requirements to the analyzed mobile installation, impose some limitations on use this or that fuel, oxidizer and ballast gas. The main difference of this installation from existing lasers is the requirement of large duration of continuous operation.

For processing of the greatest possible area of the polluted basin, the laser installation should have an operating time appropriate to operation capabilities of a vehicle, on which it is secured. For the helicopter this time makes $t \sim 30$ min. The carried estimations of interaction of a beam with a water surface covered with thin petroleum film, have shown, that the power losses by heating water can be reduces through fast scanning of a surface by a beam of power $W \sim 100$ kW. These requirements allow to estimate total of fuel and ballast gas necessary for mobile GDL operation. The specific power (W_{out}) existing homogeneous CO_2—GDL, as a rule, does not exceed 10 kJ/kg. Thus weight of fuel and ballast gas for such GDL operation should be not less than $m = Wt/W_{out} = 18{,}000$ kg. This value essentially will increase at the account of weight of the equipment for storage of working components. Apparently, that such large weight is unacceptable for laser arrangement on a flight vehicle.

At the same time, the main part (80.95 %) of weight of design components used for creation of a propulsive mass GDL, is the weight of an oxidizer and ballast gas. Thus, use of free air as an m_Σ oxidizer and a ballast gas is a unique opportunity to guaranty this installation operation.

- As CO_2—GDL fuel usually carbon monoxide is used with the hydrogen addition. Carbon monoxide allows to receive high values of specific power of radiation in comparison with liquid fuels, such as toluene, benzole, kerosene, but weight this fuel together with a system of storage is rather high. Weight CO is 13–15 % from total mass of spent components, i.e., in our case of 2,000–3,000 kg. Weight of a storage system of such quantity gaseous carbon monoxide is ≈7,000 kg, and summarized weight fuel and system of storage $m_\Sigma = 9{,}000$–10,000 kg. In case of use cryogenic state carbon monoxide summarized weight m_Σ can be reduced approximately twofold, but nevertheless it considerably surpasses appropriate weight for liquid hydrocarbon. Weight of liquid hydrocarbon is 4.5 % from total mass of spent components, and the tanks for storage are much lighter than for carbon monoxide, as the hydrocarbon are stored at atmospheric pressure. In our case weight of liquid hydrocarbon is 700–900 kg, and summarized weight fuel and system of storage $m_\Sigma = 1{,}000$–1,500 kg.
- Taking into account the mentioned above reasons, in the present operation as possible fuel are considered benzole, toluene and kerosene, and as an oxidizer and a ballast gas (instead of nitrogen) free air compressed by the compressor.
- As it was mentioned above, for normal operation of supersonic nozzles of the laser it is necessary to ensure high total pressure of a stream ($P_0 = 2$–3 MPa). The aviation compressors can ensure a compression ratio up to $P_0/P_a = 20$. Here P_0 stagnation pressure of a working mixture in a GDL channel, P_a atmospheric pressure. In all further calculations the value of stagnation pressure of a flow $P_0 = 2$ MPa and maximum values, appropriate Mach number of the nozzle M = 4.7 with the area ratio of an exit of the nozzle to the critical cross-section $F/F^* \approx 25$ was accepted.

At comparison various fuels, basic parameter is the power efficiency, namely specific power of laser radiation W_{out} (power of laser radiation referred to summarized mass flow of laser components), achievable at use of the given fuel. In this connection the analytical—theoretical research of influence of various parameters on value W_{out} was carried out.

Temperature and chemistry of a working mixture formed at combustion various fuels in air, were determined by thermodynamic calculation. The necessary for calculation enthalpy of initial components was taken from [1]. For the reason of uncertainty of efficiency of the compressor (approximately it is 0.75–0.83) in a main part of executed calculations temperature of inlet air in the combustion chamber was assumed to equal appropriate temperature at adiabatic compression up to value $P_0/P_a = 20$. This temperature calculated according to the following formula

$$T = T_a \cdot \left(\frac{P_0}{P_a}\right)^{(\gamma-1)/\gamma},$$

where T_a temperature of free air, γ—isentropic exponent. At adopted values $T_a = 298$ K, $\gamma = 1.4$ temperatures of inlet air in the combustion chamber is $T \sim 700$ K. The enthalpy, appropriate to this temperature of air is equal $I = 400$ kJ/kg.

At thermodynamic calculation the thermal losses which depend on a design of the combustion chamber and beforehand are not known were not taken into account. In this connection in the subsequent calculations of flow of a working mixture in the nozzle block inlet temperature (*K*) was taken equal to 0.95 of temperature obtained in thermodynamic calculation.

The thermodynamic calculations have shown, that the working mixture contains in the investigated temperature range (or oxidizer-to-fuel ratios) only necessary for operation CO_2—GDL components (CO_2, N_2, H_2O) and oxygen (O_2), as the content of remaining components is insignificant. Therefore in the further researches it was supposed, that the working mixture consists only of carbon dioxide, nitrogen, water and oxygen.

In the technological report (2) executed on the second phase of activities under the project 1099, the calculations of non-equilibrium flow of a mixture CO_2–N_2–H_2O–O_2 in the supersonic nozzle are carried out. The kinetic model describing population change of separate oscillating levels of molecules CO_2 and N_2 at processes of *VT*- and *VV*- of exchange was used. All oscillating levels of a molecule CO_2 with characteristic temperatures $\theta < 5{,}000$ K were taken into account. It was supposed also, that the velocity of exchange of oscillating energy to a Fermi-resonance considerably surpasses a velocity of all remaining *VT*- and *VV*- processes. Thereof, the modeling system of levels was introduced, some of which have a large degree of degeneration and correspond to several levels in a full system of levels of a molecule CO_2.

As a result of calculations the population inversion and gain coefficient of the nozzle unit outlet was determined.

The calculations of specific output power of laser radiation were conducted at following input data.

1. The working mixture of gas dynamic laser is formed at combustion hydrocarbon fuel in air. The stagnation pressure of a flow is equal 2 MPa, total temperature is equal 1,600 K.
2. The nozzle unit of the installation consists of flat contoured nozzles with the height of critical cross-section 0.8 mm and expansion ratio 25.
3. The sizes of exit cross-section of the nozzle unit: length—1,600 mm, height—130 mm.
4. The size of a cavity of the resonator along the flow direction—200 mm.
5. The three pass unstable resonator with a multiplication factor $M = 1.4$. A reflection coefficient of mirrors $r = 0.98$, the losses on dissipation make $\beta = 0.03\ \mathrm{m}^{-1}$.

The contour of a supersonic part of the nozzle was obtained as follows. The flow of gas with an isentropic exponent equal 1.35 in the contoured nozzle with an angular point was calculated and as a contour of a supersonic part of the nozzle the form of a streamline $\psi = 0.9$ was used.

The calculation results for a case of dry air are listed in Table 18.3.

Table 18.3 The calculations result of specific output power of laser radiation

Fuel	k_v (m^{-1})	w_{max} (kJ/kg)	w_{out} (kJ/kg)
Benzole	0.41	23.6	9.6
Toluene	0.39	23.1	9.2
Kerosene	0.32	20.8	7.4

Thus, value of specific output power of laser radiation, which it is possible to expect at use as fuel benzole and toluene, and as an oxidizer and ballast gas—dry air, is ≈99.5 kJ/kg. At use of kerosene this value is approximately 2 kJ/kg less.

By selection of fuel, except for power efficiency it is necessary to take into account production and toxic characteristics, and cost of components.

The application of benzole is inconvenient at low environment temperature. The freezing temperature is 5.5 °C. For toluene this value is equal −80 °C, and for kerosene −38 °C.

The limit for room vapor concentration of researched fuel have following values: 5 mg/m^3 for benzole, 50 mg/m^3 for toluene, 300 mg/m^3 for kerosene.

From said it is possible to conclude, that the application of benzole is unwise, as it a little surpasses toluene according to power efficiency, but is significantly worse for toxic than toluene and is less convenient in operation.

As to selection between toluene and kerosene, the application of toluene provides specific power increase approximately 25 %. However use of kerosene has a number of advantages. At first, kerosene is approximately three times cheaper than toluene. Secondly, kerosene is fuel for airplane and engines. Therefore at the arrangement of the laser installation working on kerosene on a flight vehicle the additional system for fuel storage is not required.

As conclusion for this chapter it is possible to note the following.

The homogeneous laser working on combustion hydrocarbon fuel products is most reasonable for development of the mobile installation.

The specific requirements to the mobile laser installation, impose the limitations on use this or that fuel, oxidizer and ballast gas. The main limiting factor is the required long duration of continuous operation. The carried out estimations have shown, that summarized weight of fuel, oxidizer and ballast gas necessary for GDL operation, is very height and is unacceptable for arrangement on a flight vehicle. At the same time, the main part of weight of fuel components used for GDL operation mass, is weight of an oxidizer and a ballast gas. In this connection, use of free air compressed by the aviation compressor, as an oxidizer and a ballast gas is the only way to organized operation of this installation.

The carried out comparative analysis of various fuel has shown, that most reasonable among them are toluene and kerosene. The application of toluene provides specific power of the installation approximately 25 % higher than that of kerosene. However, use of kerosene has a number of the advantages. At first, kerosene is approximately three times cheaper than toluene. Secondly, kerosene is the fuel for airplane and helicopter. Therefore at arrangement of the kerosene operating installation, on a flight vehicle the additional system for storage fuel is not required.

18.4 Selection of the Power Installation for GDL

18.4.1 Selection of a Schematic of the Power Installation

In the previous part devoted to the computational substantiation of the basic power characteristics, the capability of development of the mobile CO_2—GDL with use of kerosene as fuel, and as an oxidizer and a ballast gas—atmospheric air supplied by the compressor is shown.

Besides, some GDL parameters, permitting to determine the required characteristics of the compressor are added.

In particular:

The calculations have shown, that in GDL should be realized the condition:

Po /Ph = 20

where:

Po stagnation pressure of a working mix(mixture) in GDL channel;
PH atmospheric pressure.

In view of channel losses from the compressor to GDL inlet of the chosen compressor should have the pressure ratio not less than 22–23.

Temperature of inlet air in GDL combustion chamber chosen at calculations is 720 K, that corresponds the temperature range of compressed air for compressors with

$$\prod \kappa > 22 \div 23.$$

For an estimation of necessary value of airflow rate through GDL it is rational to set specific energy extraction at known accepted power range of 100 kW. As was shown in the previous chapter, the GDL specific power with combustion of kerosene in air is 7.5 kJ/kg. However this value can be reached at dry air only. At increase of humidity of atmospheric air the specific power begins to decrease and at humidity value 100 % (T = 25 °C) it is 5.4 kJ/kg.

To achieve the higher specific power it is possible in principle to include air dryer in an air flow path, however it large overall dimensions and power consumption can result in technical and operational problems and, besides the automatic control of dryer operation is necessary if air humidity and fuel have changed. It will be more conveniently to have certain losses of specific power at humid air, having compensated losses by increased air flow.

Therefore, having accepted average specific power of 6 kJ/kg, instead of 7.5 kJ/kg we shall receive an estimated value of an air flow—16.25 kg/s.

Thus, for a operation of the laser installation in power range of 100 kW it is necessary to have the compressor with an air flow rate ≈16 kg/s and $\Pi\kappa > 22 \div 23$.

In essence task of air compression at flow rate of 16.25 kg/s and pressure in access of 20 atm does not contain unsolved technological problems, since compressor engineering is widely applied in industry. The task consists to find the optimal solution ensuring the reasonable cost of the installation, reliability of its activity and convenience in operation.

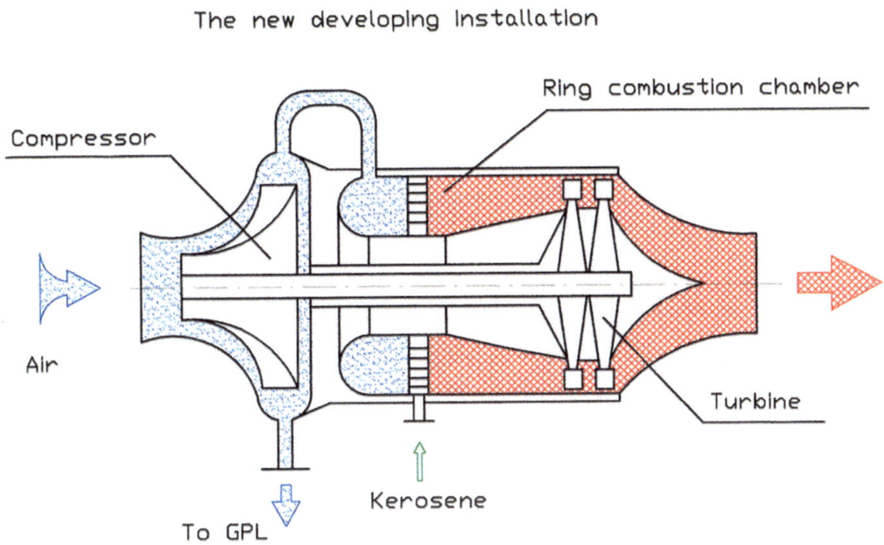

Fig. 18.5 Power installation for laser action

Let's consider three possible schematics of power installation.

18.4.1.1 Development of Special Power Installation for the Above Mentioned Parameters of the Compressor

In this case the installation will consist (Fig. 18.5) of the compressor itself, turbine for driving, gas-generator for rotation the turbine, and also fuel supply system, control and regulation units.

As there is no installation for the mentioned parameters in industry, it is necessary to develop such installation from very beginning, to conduct preproduction activity and to find a producer. For all reason, the price of the produced installation will be high, and development—time consuming.

Besides, short order book for such installations (single production) will lead initially to unprofitable operation for any producer.

18.4.1.2 Air Compressor on the Base of Aviation Engine Shaft

In general approach, the task can be decided rather simple. On the shaft of the aviation gas-turbine engine the compressor with the required characteristics is installed instead of the propeller.

The calculations show, that the shaft horsepower about 5.5–8 MW is necessary for compressor driving with the flow rate of 16 kg/s and Pк 23 kg/cm^2. It is power of the aviation gas-turbine engine of middle power class, a lot of which is in series production.

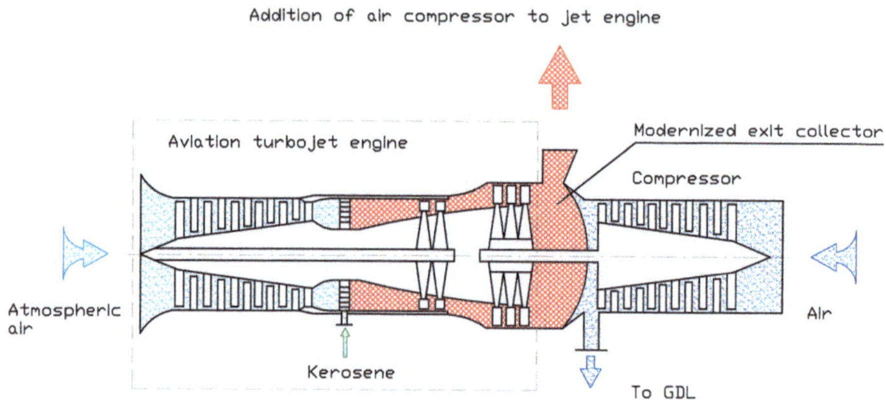

Fig. 18.6 Installation of an axial multistage compressor on the shaft of the engine

However, as engineering realization the task has a number of serious problems.

The nominal speed of the propeller shaft of existing aviation engines is in 5–8 thousand rpm range, and all high efficiency centrifugal compressors require operation speed of the order above.

For example, industry compressor TB7-117 being far from the best in efficiency already requires 30,000 rpm.

The installation between the gas-turbine engine and compressor of the high-speed gear reducer (multiplicator) because of complexity in operation and large overall dimensions excludes such solution as acceptable.

Technically task is solved by the installation on the shaft of the engine of an axial multistage compressor (Fig. 18.6). But even the principal diagram general view demonstrates the technical irrationality of such solution, it practically two aero-engines, one of which a production one, and the second requires large design and manufacturing costs connected refurbishment for realization of rigidity of the shaft, organization of turbine side exhaust etc. Besides the overall dimensions and accordingly of weight of the installation are considerably increased.

18.4.1.3 Air Bleed from Aero-engine Compressor Final Stage

As it is visible from the scheme Fig. 18.7, if it is decided to take an air bleed for GDL after a final stage of the engine compressor, practically there is no necessity redesign the power installation with installing new aggregates. Moreover, the dismantling from the engine of the free running turbine operating for driving the propeller shaft for the reason of propeller nonuse, considerably simplifies a design of the installation as a whole.

This solution bribes by the simplicity, since the engines not only will be used with minimum modification, but also all control systems are kept, including the use of the control panel which is taken off from a flight vehicle.

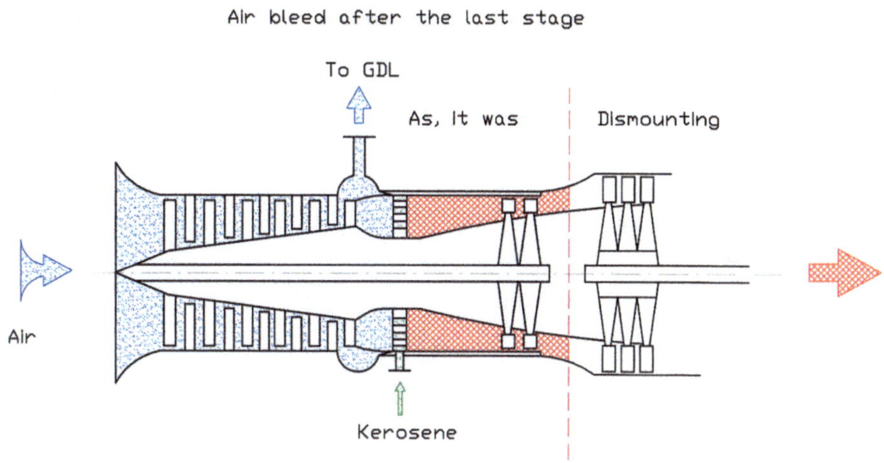

Fig. 18.7 An air bleed for GDL after a final stage of the engine compressor

Besides, from point of view of the power, dismantling of the free running turbine together with nozzle block allows to lower required power of the turbine of system of the compressor—free running turbine.

The decrease of required power of the system offers the possibility to reduce quantity of required air and kerosene in the combustion chamber of the gas-turbine engine. With constant power rate of the turbine for the drive of the compressor and of the compressor flow rate, probably, it exist the possibility to take off a part of the air and to direct it to the combustion chamber of gas dynamic laser. At the same time, utilizing the characteristics of the combustion products down the main turbine (ρ, m, T), it is possible to organize an ejection of GDL gases down the supersonic diffuser for improvement of the gas dynamic laser overall performance, especially at start.

18.4.2 Selection of the Aviation Gas-Turbine Engine

With the purpose of optimal selection of an aviation-engine for this task it is necessary to formulate the initial requirements.

1. For obtaining air pressure in GDL inlet not less than 20 atm and in view of losses in a flow channel the compressor of the engine $\Pi\kappa$ should be in range of 22–25, not less.
2. The GDL inlet air flow rate is 16.25 kg/s.
   As for elimination of extreme engine power reduction the air bleeding from the compressor should not exceed 15–18 %, the total air flow rate through the engine should be not less than 100 kg/s.

$$\lambda a = \frac{Wa}{a_*}$$

Table 18.4 The data of some Russian engines characteristics: Пк and G of airflow rate (kg/s)

Engine	НК-12СТ	НК-14СТ	НК-14Э	НК-16СТ	НК-17	НК-36СТ	ПС-90А	РД-33
Пк	8.8	9.5	9.5	9.68	9.68	23.12	19.6	22.0
G (кг/сек)	56.0	37.1	39.0	102.0	102.0	101.4	56.0	80.0

Fig. 18.8 Gas turbine engine NK-36ST

3. The chosen aviation engine should widely be used by standard aviation operation, that enables to utilize engines which are decommissioned from flight vehicles, having passed prescribed operational life limit, but still are operable for ground installation. To buy them is much cheaper, than to buy new engines.
4. It is desirable to make a choice of domestic (Russian made) engines that enables in case of realization of the project to connect to activities, both engine developing company and manufacturer.

The table data of some Russian engines characteristics: Пк and G of airflow rate (kg/s) are given in Table 18.4.

The analysis of the table results in a conclusion, that most reasonable for the laser installation could become the engine НК-36СТ serially produced by Samara plant "Motorostroitel", which satisfies to all requirements presented above. It is obvious, this is not the single engine, which can be put to use for this task, but in advantage of it would be desirable to note the following.

The high performance engine НК-36СТ of the Samara Technological complex "Motorostroitel," is developed in 1990 on the basis of an aero-engine НК-321 and is designed for the drive of the centrifugal supercharger in a structure gas pipeline pumping aggregate.

The modular design of the engine facilitates transport and assembly. The engine has the remote control panel and record of long life ground operation.

The general view of the engine is shown in Fig. 18.8. In Fig. 18.9 the design concept is shown.

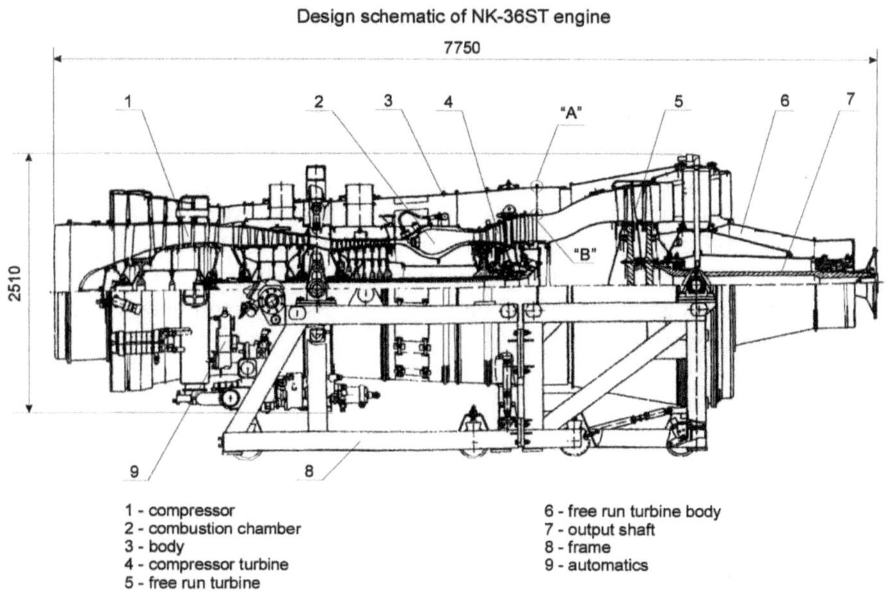

Fig. 18.9 Design schematic of NK-36ST engine

18.4.3 Gas Dynamic Assesment of the Chosen Aero-engine as the Power Unit for Mobile GDL

In the previous part the task of GDL supply with oxidizer (air) by use of the HK-36CT engine compressor was considered. However one more function—use of engine exhaust gas as working medium for ejection of an GDL exhaust system is planed to the chosen engine.

The practice of development GDL with direct exhaust of combustion products in atmosphere has shown, that the aft diffuser of the laser produced the pressure recovery of gases up to atmospheric pressure requires for start significant the combustion chamber pressure increase (more than 30 atm), or decrease of exit pressure below atmospheric one. It is possible to ensure the last requirement, through installation of an ejector, for which the exhaust gases down the engine turbine are the working medium.

Gas dynamic estimation of the capability of the HK-36CT engine of GDL diffuser start support is shown below.

As it was already indicated, the basic adaptation of the HK-36CT gas-turbine engine is the elimination of design structure of the free running turbine, that allows to improve characteristics from the point of view of possible application in the laser installation, for which it is necessary to bleed air with the consumption $m_B = 16.25$ kg/s by pressure $P \approx 22$ kg/cm^2 and temperature $T = 720$ K.

Reasonable temperature for a GDL gas flow channel is $T = 1{,}600$ K at combustion-chamber pressure of 20 kg/cm^2.

The given temperature is realized in the combustion chamber for fuel kerosene—air at mass ratio Kκ = 38.5, that is at kerosene flow rate $m_n = 0.422$ kg/s.

In GDL working section it is necessary to ensure static temperature of combustion products Ta = 370 K, that is realized velocity exhaust of combustion products through supersonic nozzles with velocity Wà = 1,700 m/s at a Mach number Ma = 4.46, isoentropy index κ = 1.333, pressure Pa = 0.057 kg/cm^2.

For realization in GDL flat working section shock wave free flow of the of combustion products in atmosphere is necessary to utilize the flat aft supersonic diffuser. One of major parameters of functioning of the diffuser in GDL operation conditions is the start pressure.The start pressure (Pst) is determined on intensity of a direct shock wave relation:

$$\frac{Pst.}{P_{\text{H}}} = \frac{g\left(\frac{1}{\lambda a}\right)}{g(\lambda a)};$$

where:

PH	pressure behind the diffuser;
$\lambda a = \frac{Wa}{a_*}$	velocity factor equal to the ratio of exhaust velocity in working section GDL (Wa) to a critical velocity of a sound (a*).

The value of a velocity factor λa is connected to a Mach number by a ratio

$$\lambda = \sqrt{\frac{X+1}{2} \cdot \frac{Ma}{\sqrt{1 + \frac{X-1}{2} Ma^2}}}$$

and then to a Mach number Ma = 4.46 corresponds $\lambda a = 2.32$.

The function $g = \frac{F\kappa p}{Fa}$ determines the ratio of the geometrical area of critical cross-section of the diffuser to the area of gases exhaust and is according to expression

$$g(\lambda a) = (\frac{X+1}{2})\frac{1}{X-1} \cdot \lambda a(1 - \frac{X-1}{X+1}\lambda a)\frac{1}{X-1}$$

For direct shock wave the following ratio is valid

$$\lambda a \cdot \lambda_{n.c.} = 1;$$

where:

$\lambda_{n.c.}$ dimensionless velocity factor after front of a direct shock wave

Thus, the ratio of pressure of start to pressure behind the diffuser is equal

$$\frac{P_3 an.}{P_H} = 13.7.$$

The wind tunnel pressure of start is usually above their steady state operating pressure up to 10–30 %.

At the same time, in GDL laser conditions, when there is a involved gas volume in a working section (region of resonator mirrors) pressure of start of the diffuser can is still increasing to the value, at which the diffuser could not started.

Hence, it is necessary to lower exhaust pressure from the diffuser, that is possible to realized through an ejection of a gas jet, flowing out from the diffuser, by a gas jet flowing out after the turbine of the engine (Fig. 18.10).

The designs of similar ejectors with maximum pressure recovery are known at high speeds of flow of gas in minimum cross-section. Such ejectors designs are characterized by smooth profile of an inlet part and with small cone angles (L) on an exit, which make value L = 6°–10°.

The pressure recovery coefficient for such ejectors can reach value

$$\frac{Pexit}{Pinlet} = 0.955 \div 0.965 \text{ and higher.}$$

Thus the velocity factor value in minimum cross-section should is

$$\lambda = 0.75 - 0.85.$$

Thus such ejector inlet pressure of should be Pinlet. = 1.076 kg/cm^2 (Pexit. = 1.033 kg/cm^2), that can be provided with the gas-turbine engine without the free running turbine.

At a velocity factor $\lambda = 0.8$ in minimum cross-section and ejector inlet pressure Pinlet. = 1.076 kg/cm^2 the static pressure in minimum cross-section becomes equal Pmin = 0.73 kg/cm^2, that allows to organize introduction of flowing out

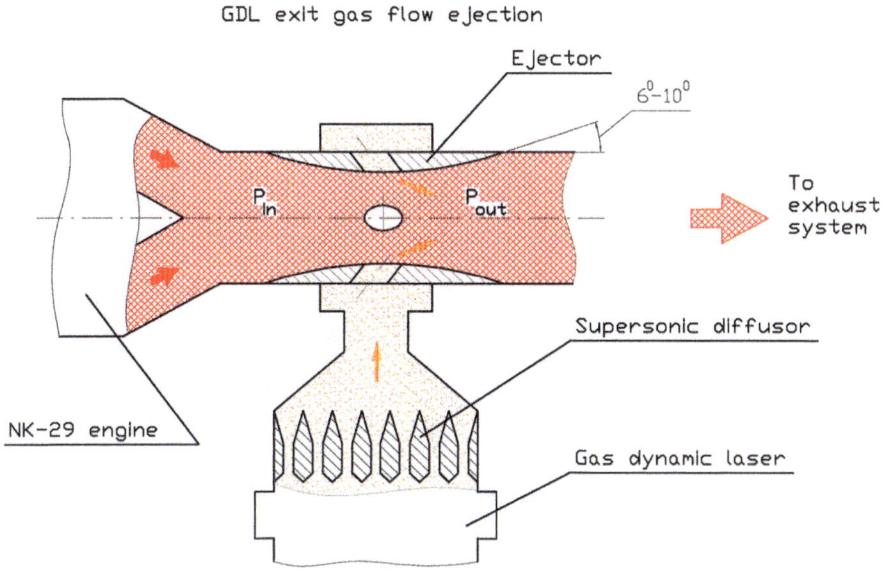

Fig. 18.10 GDL exit gas flow ejection

from the GDL diffuser gas in area of minimum cross-section of the exhaust device of the gas-turbine engine and by that to lower pressure behind the diffuser GDL at least down to Pexit = 0.8 instead of 1.033.

Such reduction of pressure will allow to lower pressure of start GDL down to a reasonable level of 18–20 atm.

18.4.4 Design Solution for GDL Power Unit on the Basis of the Gas-Turbine Engine

In view of the technical plan generated by the authors of the present technical proposal, experts of the Samara complex “Motorostroitel” , has developed on the basis of the HK-36CT gas-turbine engine the engine for a laser complex under index HK-29.

The HK-29 engine (Fig. 18.11) represents the gas-turbine engine obtained as a result of adaptation of the production engine HK-36CT, which consists in following:

18.4.4.1 Dismantling of the Free Running Turbine with the Output Shaft

The dismantling is made through the joint “A” flange of a engine body (see Fig. 4), through the joint “B” of the hot gas pipe down the turbine of the compressor and finally the free running turbine together with a turbine casing is removed from the engine. As the frame, on which the engine is mounted, consists of two not

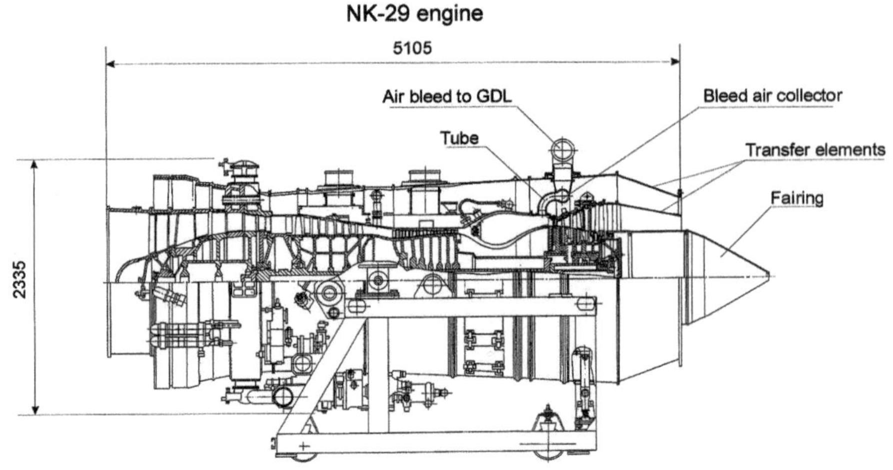

Fig. 18.11 NK-29 engine

connected permanently sections, the removal of the second, turbine side section the does not represent complexity.

For organization of the smooth gas flow from the modified engine, and also for a fixing of the exhaust system to the engine, two conical adapters (casing) and internal fairing made of a stainless steel sheet are added.

The external adapter is terminated by a flange to be fixed with the engine ejector. The adapters and fairing were designed by "Motorostroitel", supplier in accordance to the aerodynamic characteristics of a gas stream and the strength requirements.

Ejector, which scheme is shown in Fig. 18.10, is not the engine item. The ejector design is calculated and finally developed on the basis of engine, and laser complex exhaust system parameters and GDL exit characteristics (down the supersonic diffuser).

Ejector design represents a pipe from high-temperature steel having symmetrical narrowing in the middle part ahead of critical cross-section. The pipe can be manufactured from three parts with flange or welded connections.

The middle narrowing part can be turned from a thick-walled billet or from a rolled ring and has mating collars for different side collector for convenience of assembly for welding. The orifices connecting cavities of exhausts of the engine and GDL are located around the perimeter of critical cross-section. The collector is manufactured by stamping of a sheet, has the welded branch pipe and serves for uniform distribution of gas after GDL on all cross-section of an ejector. Overall dimensions of an ejector and final design will be submitted after fulfillment of general layout of laser complex in the final report.

The adaptation of the engine with removal of the free running turbine considerably reduces overall dimensions and weight of the installation. For a clarity of estimation of advantage in overall dimension reduction of the modified engine the presentation on Figs. 18.9 and 18.11 is made in the same scale.

18.4.4.2 Air Bleeding to GDL

The next considered problem is the air bleeding after the compressor to GDL

During searching of the engineering solution for the engine adaptation the aviation specialists has offered to realize the air bleeding from the exit of the combustion chamber second contour, not from the compressor final stage exit, i.e., from a channel of a cooling flow of the combustion chamber. This solution is connected to the fact, that combustion chamber of the gas-turbine engine is one of the most thermal stressed items of the engine, and the cooling parameters are results of long development stage. Therefore the air bleed behind the compressor, reducing the air flow of the combustion chamber cooling flow, will change steady state thermal status, that can result in undesirable consequences for engine operation.

The collector ring for air bleeding with a branch pipe to GDL supply was recommended to position in a zone of the exit of cooling flow of the combustion chamber, and for that the chamber and collector are connected by angular branch pipes (Fig. 18.12).

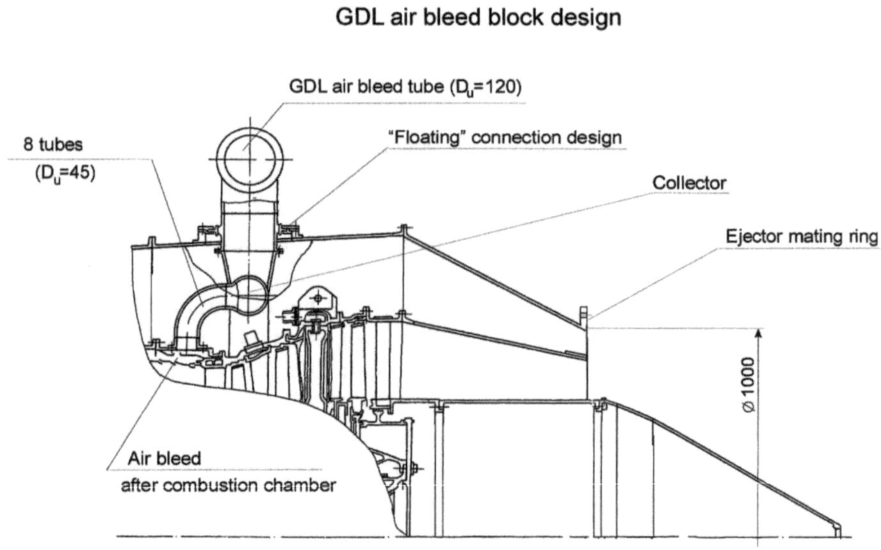

Fig. 18.12 GDL air bleed block design

The angular branch pipes, eight pieces in quantity, can be manufactured of a pipe or stamped sheet from two halves and have leak tight flanges to cooperate with the envelope of chamber cooling flow channel. The angular branch pipes are fixed by welding to the collector ring, which has the terminal flanged tap. To this tap the branch pipe to supply air to GDL is fixed. This branch pipe passes through an outer shell of a engine body and has "floating" attachment to it excluding deformation of designs at a thermal expansion of contacting elements.

It exists the experience of such bleed schema, however particular design should pass a number of development tests with the purpose of experimental selection of channel pressure reduction to ensure calculated bleed air flow rate.

All other elements of the engine remain without change, including all start and cutoff automatics and operation control.

Thus engine НК-29, obtained as a result of adaptation of the engine НК-36CT will require minimum development, and manufacturing costs.

The basic performances of the engine НК-29 are given in Table 18.5.

All parameters of the engine are indicated for a nominal mode.

The engine НК-29 is single mode, as by activity of the laser installation is required neither decrease, nor increase of the power, that provides reliability and life value growth.

The engine НК-29 should be manufactured by aviation plant irrespective of, whether it is refurbished through adaptation of the engine НК-36CT or again assembled with use of furnished engine parts from production engines.

Thus, as a result of the carried out research work not only the optimistic results for possibility of adaptation of the gas-turbine engine for GDL gas dynamic

Table 18.5 Main parameters of HK-29 engine

Fuel	Kerosene TC-1
Power (MW)	25
Air flow (kg/s)	107.2
Compressor pressure ratio (Пк)	23.12
CC gas temperature (K)	1,420
Gas exit temperature (K)	698
Propellant consumption (kg/s)	1.434
Engine weight, with frames (kg)	~5,000

complex were obtained, but also design activities by definition of general realization of such engine are executed in practice. And, what is very valuable, this activity is executed by the experts of aero-engines developing—producing company.

18.5 The Mobile Laser Installation

18.5.1 The Pneumahydraulic Scheme of Laser Installation

PHS of the laser was developed from the preliminary conditions:

- Assurance of required calculated parameters in all PHS elements;
- Automatic control of physical processes;
- Use of minimum set of commands and control units;
- Design simplicity of the scheme and safety in operation.

PHS is shown in a Fig. 18.13. It includes directly gas channel—gasgenerator (GG), nozzle unit (NU), working section, supersonic diffuser and exhaust line.

Besides, the GDL air-supply system, consisting from supply line, air valve and disk pressure reducer; the fuel supply system including pipelines, fuel pump, valves and adjusting flow restrictors are included into PHS. To PHS the aerodynamic window also included.

Shown on the scheme the optical resonator and protective shutters are not PHS components, but their functions are interconnected and they participate in operation process.

For automatic control of all cycle of activity GDL (start, steady state mode and shut down) and parameter monitoring, PHS is equipped with the measuring equipment producing signals through electronic units on functional control units of an automatics and on the operator console.

The measuring and control wiring schematics are not added in the report, as they do not require special development for the reason of ordinary professional level.

The required electrical voltage is 27 volts and power source—the electric generator of the aviation engine.

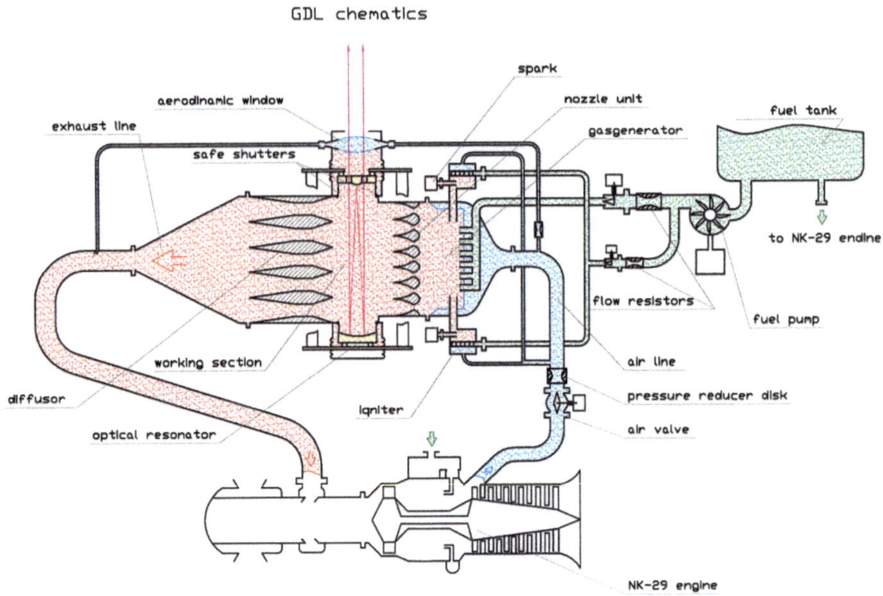

Fig. 18.13 GDL schematics

Start of the laser installation

The PHS element operation begins with opening of the air valve with pressing by the operator on the button “start” at aviation engine (further engine) being in operation status.

The valve opens for delivery half of air flow and hot air is supplied in the gas-generator and igniters.

The command to open valve is a start command for GDL operation, all following commands to operation of elements of the scheme are made automatically, without interference of the operator.

As compressed air pressure is increased to 10 kg/cm^2 the pressure transducer gives a signal, according to which the electrical voltage is supplied to spark plugs of spark igniters and for the electric drive of a fuel pump GG. The valves of GG and spark igniter fuel supply are closed.

The lead of voltage application to spark igniter before fuel supply operation excludes a capability explosive combustion during start process.

At 20 kg/cm^2 fuel pump exit pressure, the command is given for opening spark igniter fuel supply valve. The high pressure fuel supply guarantees the high quality fuel atomization in injector, thus reliable ignition. Igniters begin functioning, forming “forced” plumes directed against each other. The igniter plumes heats up sensing elements of temperature detectors installed in bodies of prechambers and on reaching of 950 K temperature the signal is given on opening of the main fuel valve.

Fuel is supplied to the GG injector first row and being dispersed forms with air a fine mixture, which is ignited by spark igniter plumes. The complete burning of fuel has taken place on the injector second row. GG temperature increases and at reaching value 1,050–1,100 K the commands are given to full opening of the air valve and valve of air delivery in the aerodynamic window.

The generator is forced to a steady state mode through idle transfer mode. The working gas is accelerated in the nozzle block up to a velocity M = 4, and after the working section (the resonator zone) is decelerated in the supersonic diffuser and downstream through an exhaust collector leaves the gas channel of the laser. The gas dynamic channel flow is stabilized, providing along all length of a working section static pressure equal 0.12 kg/cm^2. The vacuum sensors give a signal on opening of protective shutters and spark igniter cutoff (closing of the fuel valve and deenergizing spark igniter plugs). The air flow through switched off igniters protects the igniter plugs from an overheating during all operation cycle.

The opening of mirrors by shutters the optical resonator begins to form in an active medium of a working section an electromagnetic laser field with extraction of radiation. GDL is functioning.

GDL steady state mode provide a constant pressure of gas at the engine compressor exit and that at fuel pump exit, as well as constant hydraulic resistance of air pressure reducing disks and fuel flow restrictor.

GDL is equipped with a system of emergency protection, which on an unacceptable falling or overshot GG temperature closes the fuel valve instantly, switches off a fuel pump and forms a command on the engine shutoff.

The GDL emergency shutoff can be executed by operator pressing the button "emergency shutoff", with similar command sequence on actuators.

18.5.1.1 The Normal Cutoff of the Laser Installation Happens in the Following Order

By operator pressing the button "shutdown", protective shutters and the GG fuel supply valve are closed, and the fuel pump is switched off. The supply GG fuel ceases and during 7–10 s gas channel purging by pure air has taken place, then the command on closing of the air valve is given and the engine is shutdown.

The GDL cycle sequence of start and shutdown is shown in Fig. 18.14.

As it can be seen according the cyclogramme, each next command on functioning of the aggregate is given after fixing the results from the previous command. For this reason, at a presentation of these results on the display of the operator it is possible to control the start sequence manually, having disconnected an automatics of control.

The ranges on time between commands and their fulfilment are determined at designer operational development GDL.

18.5.2 GDL—General Design Features

On the basis of computational examinations presented in the previous reports, the GDL general design features is generated according to recommended pneumohydraulic scheme.

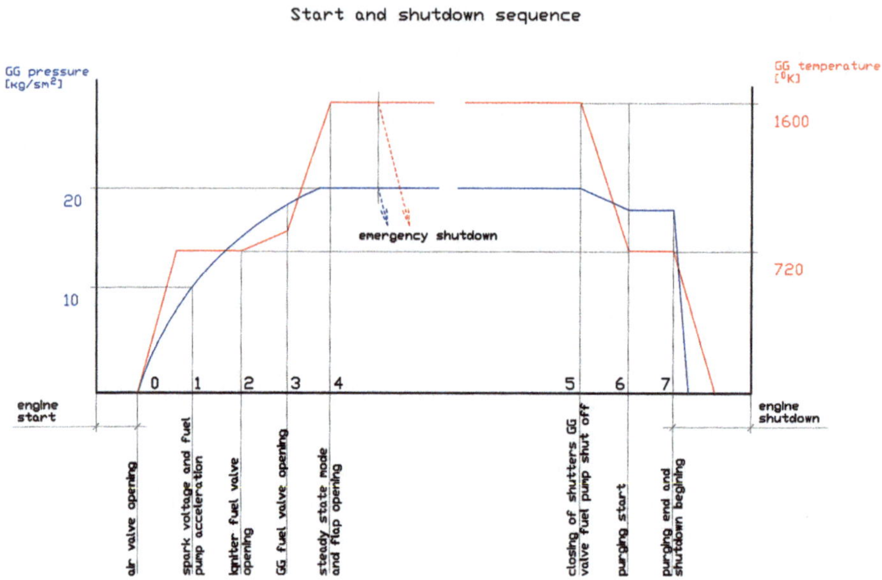

Fig. 18.14 Start and shutdown sequence

At GDL designing authors tried to decide a task of the greatest possible simplification of a design with the purpose to cut cost (in general GDL is in category of expensive products).

For this purpose, firstly the single mode scheme of the laser excluding necessity to refurnish of the laser with a complex control system, was chosen. Secondly, the completely uncooled design despite on a rather high operation temperature of gases is offered.

The last solution is stipulated by the following:

(a) Absence mobile and the rotating parts in GDL design—there is no necessity to use greasing materials;
(b) Absence of a strict limitation on weight, and, hence, on thickness of walls and weight of details;
(c) Capability to limit the duration of continuous operation;
(d) Presence on the market of heat resisting structural materials, beginning from special alloys to structure ceramics.

The GDL general view is shown on Fig. 18.15.

GDL units are the following:

1. Gasgenerator (GG) with spark igniters.
2. Nozzle unit (NU) with a working section (gas part of the resonator).
3. Supersonic diffuser.
4. Optical resonator with mirrors.
5. Chamber of the resonator.
6. Aerodynamic window.

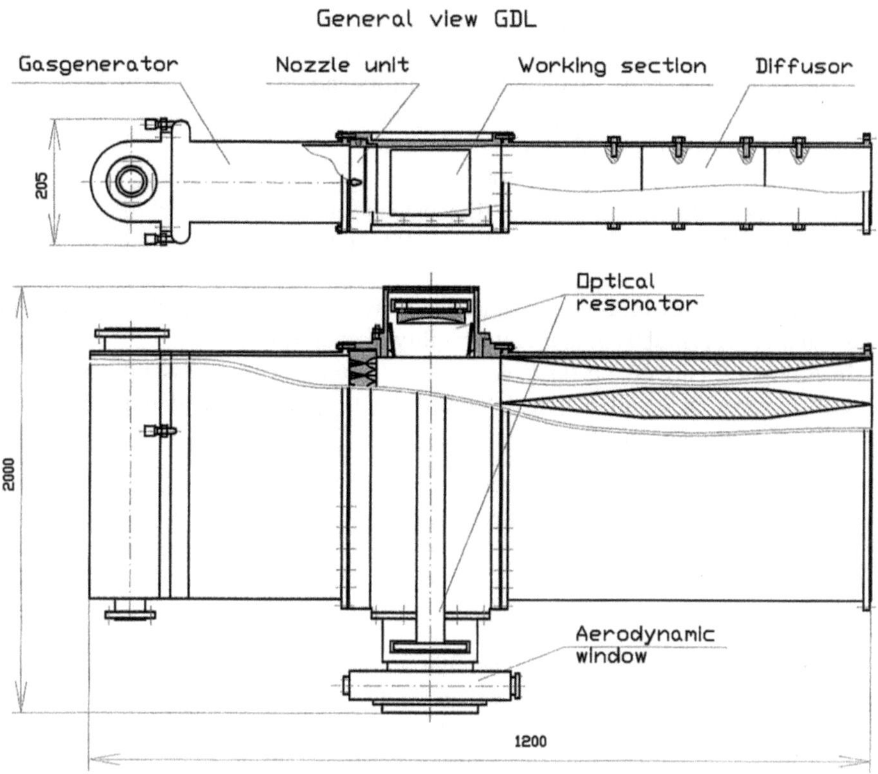

Fig. 18.15 General view GDL

7. Fuel pump.
8. Air valve.
9. Fuel valve (2 items).
10. Component supply units.
11. Attachment units.
12. Measurement system sensors.
13. Cable configuration.

For the base sizes determining all remaining overall critical dimensions of main assemblies, the sizes of cross section of the nozzle unit were chosen.

According to the formula (2) (report # 2), at the summarized flow rate of working gas 16.647 kg/s (air—16.25 kg/s, kerosene—0.422 kg/s), temperature—1,600 °K and pressure of 20 kg/cm^2 we obtain the area of critical cross-section F* = 8,524 mm^2.

At adopted nozzle area ratio

$$F/F^* = 25$$

$$F = 213{,}100 \text{ mm}^2.$$

Ratio between the height of a nozzle blade and width of the nozzle unit (across a flow) was chosen from experience of improvement of gas dynamic lasers and rational relationship between diameter of mirrors and distance between them.

In opinion of the authors, optimal values for the obtained area are:

height of a blade H = 130 mm;

width of the nozzle unit L = 1,600 mm.

These sizes determine overall dimensions of the generator of gases and supersonic diffuser coupled with the nozzle unit, and, hence all GDL design coordination.

All mating places of aggregates are made as detachable flanged coupling with seals, that provides GDL high maintainability at operation.

In following part of the book the design features of main assemblies will be described more in detail which determine GDL specificity and require design-technology improvement. They are:

- The generator of gases with igniter;
- The nozzle unit with a working section;
- The supersonic diffuser;
- The aerodynamic window;
- The optical resonator.

All remaining elements of the scheme GDL do not represent engineering complexity and do not require special designer studies. In general, these elements will be utilized as serially produced by other industry branches (fuel pump, valves, air valve, the measuring equipment, the electric drives) or those are projected the elementary assemblies, which design is obvious and is determined by their assigning. They are:

(a) Propellant component supply units—pipelines of a necessary configuration and diameter, branch pipes, flow rate controlling orifices and pressure reducers
(b) Attachment units—general frame, mounting brackets of aggregates, plates for attachment of sensors etc.
(c) Resonator chamber—usual casings shielding the mirrors and ensuring air-tightness of the resonator cavity. The chamber is fixed to flanges of the working section and is connected to racks of the resonator through elastic elements. The chamber has removable covers for an optical adjustment of the resonator, one of covers (an output mirror chamber) has a window docked with the gas dynamic window.
(d) Protective shutters—(Fig. 4) two-folding type rotated around of axis for opening and closing, located between windows of a working section and resonator chambers. Basically shutters can be of any design and can be fasten or on a flange of a working section, or on a flange of the resonator chamber. In this case their attachment to flanges of resonator chambers is supposed.
(e) Adapter—part of a gas line for flow transfer and modification from rectangular cross-section down the diffuser to round cross-section for mating with GDL exhaust pipe.
(f) Multicore wire cable having wire for control and measurement transmission. The cable has thermal insulation and is fixed on the installation.

GDL is assembled on the frame, which through rubber anti-vibration mountings is mounted on a separate platform. The shape of the platform and fixing to a vehicle places are determined under design condition of this vehicle.

18.5.3 Main GDL Assemblies

18.5.3.1 Assigning and Design

In the given chapter the conceptual approach to designing of GDL main assemblies is considered. The present operation on research of a capability of GDL development for the solution of a specific task did not assumed detail design-technology study of parts, assembles and installation as a whole, however authors made initial studies of main assemblies with the purpose to show the reality of their development, definition of structural and technological peculiarities, and estimation of their dimensional and mass characteristics.

18.5.3.2 Gasgenerator (GG)

The generator of gases is intended for complete (without soot forming) burning of kerosene in hot air for obtaining a working mixture of required of a chemistry, temperature and pressure. For uniform of all mentioned parameter flow at nozzle unit inlet GG is configured of rectangular cross-section identical to cross-section of the nozzle unit.

The philosophy of designing of combustion chambers of aviation gas-turbine engines having good technical tradition and wide experience of operation was included in GG design.

All aviation gas-turbine engines combustion chambers have a body (cold part) and heat jacket (hot part). The entry of air is organized through injectors and orifices in the heat jacket located on all of its surface.

The similar GG design is offered by the authors (see Fig. 18.16).

GG consists of body of the rectangular shape, heat jacket, burner plate, air manifold and two igniters.

The body has two flanges for mating with the nozzle unit and air manifold and inside has mating surface for the heat jacket with a burner plate. On lateral walls of the body there are orifices with flanges for the igniter installation. From above and from below to a body fuel collectors are welded with exit pipe connections directed inside of bodies to feed the injectors.

The burner plate has two walls with the installation sites for 28 injectors located in two rows. The first wall (flow down direction) has orifices for air entry in intermediate space. Air through orifices in injectors reaches the atomization zone.

The pipe connections of injectors are connected to pipe connections of a fuel manifold by tubes with nipple joints. The tubes are bent for compensation of

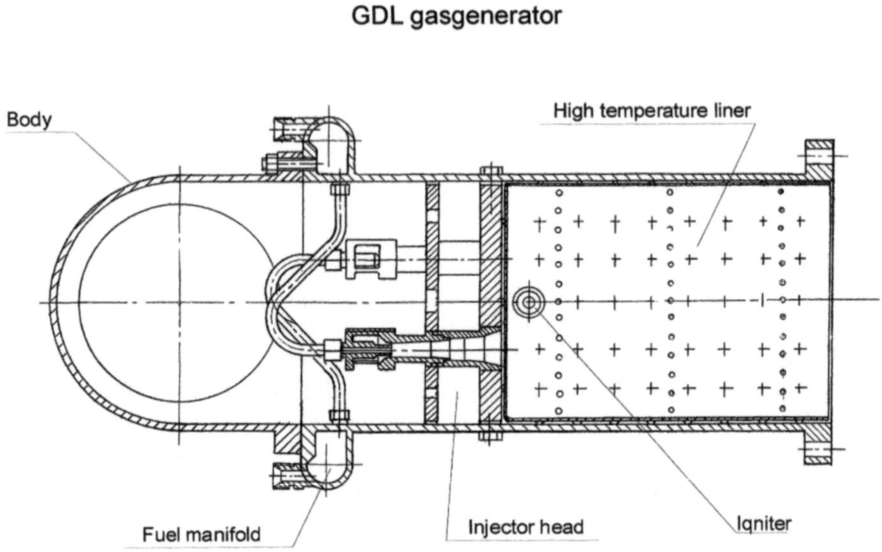

Fig. 18.16 GDL gasgenerator

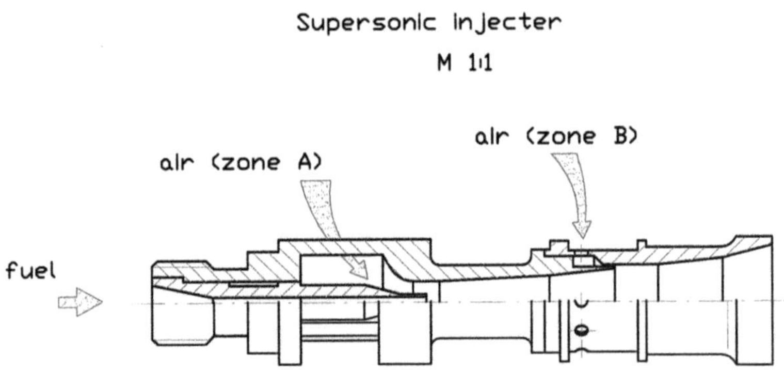

Fig. 18.17 Supersonic injector

thermal stresses, and their length is increased at the expense of connection of the lower row of injectors with the upper fuel manifold and on the contrary for same purpose.

New GG design feature is the application of supersonic injectors. In Fig. 18.17 in 1:1 scale the design of such injector is shown. The total number of injectors is 28 with 14 pieces in row. The injector has two mixing zones of fuel with air. The first zone (A) represents a Laval nozzle where air accelerates to the speed of sound with the introduction of fuel in the region of critical cross-section, the second zone (B) used for pressing of fuel drops of from walls of the nozzle and increases mixing efficiency. Air in the second zone enters from intermediate cavity of a burner plate. This injector design is patented and already have found broad application in various power plants

The air flow rate through injectors is 6.25 kg/s, what corresponds $\alpha \approx 1$.

The remaining air flow (10 kg/s) is directed through orifices in the heat jacket. Jacket is made of a thin-walled heat resisting sheet and is welded by spot welding to the burner plate. At calculated value of a residence time of mixture length of GG "hot" part is chosen 180 mm. In a heat jacket three rows of low pressure reduction orifices are made, which diameter will be updated during GG experimental improvement.

The GG gas collector is the adapter from a round flange of inner diameter = 150 mm to GG rectangular cross-section.

Igniters are introduced into a design for increase of reliability of mixture ignition as reliable ignition is problematic at the large flow rates and velocities of gas with direct ignition by spark igniters.

The igniter design is traditional, it consists of body having air inlet, one centrifugal fuel injector, electrical spark for ignition and prechamber with the nozzle.

The pressure difference along the nozzle provides necessary length of a plume in GG.

On GG body there is the position place for pressure and temperature transducers, and on a prechamber—for temperature detector.

18.5.3.3 Nozzle Unit (NU)

NU provides acceleration of a working mixture up to supersonic velocities ($M > 4$) for gas temperature reduction down to value 300–350 K, ensuring obtaining, population inversion of laser levels of a molecule CO_2.

In connection with that the working section structurally is the part of NU structure, the NU second function is the maintenance of the supersonic uniform expansion of working mixture in resonator zone.

Structurally NU (Fig. 18.18) consists of body, complete set of nozzle blades and elements of attachment of nozzle blades in the body.

The body is made as one unit with a working section, has a mating place for the set of nozzle blades and transfers in working section of length 200 mm. With the purpose of gas dynamic losses reduction on walls, which can result in deceleration of a supersonic flow a top and bottom wall of a working section extend down streamwise at angle $+3°$.

Thus, the cross-section of a flow down blades 130×1600 transfers in the flow cross-section in front of the diffuser 142×1600. In NU body in a zone of a working section there are places of vacuum sensors installation.

The NU supersonic channels are formed by profiles of two neighbor nozzle blades (Fig. 18.19a). The sizes of a channel, and, hence, geometry of blades were chosen by results of calculations presented in the previous reports and in view of structurally—technological peculiarities of blades.

In particular size of critical cross-section is chosen (according to experience) $h_{cr} = 0.8$ mm. At the smaller sizes h_{cr} the laser characteristics are improved,

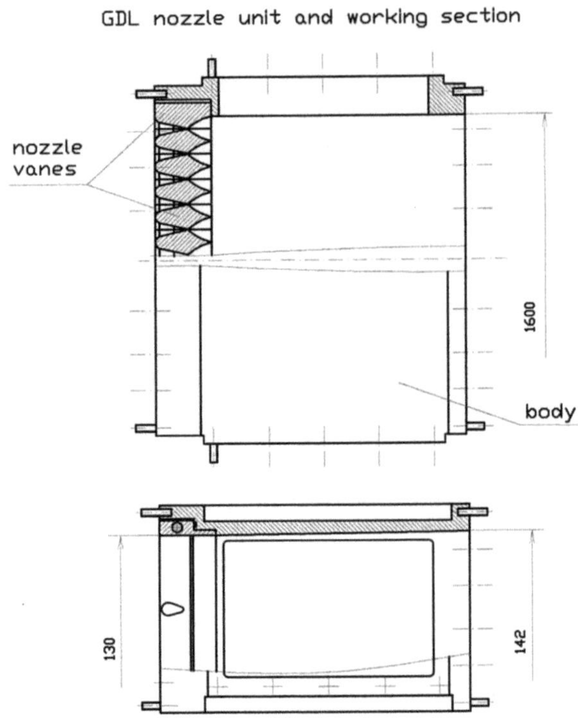

Fig. 18.18 GDL nozzle unit and working section

however blades technology factors are poorer and the error of manufacturing of blades on the size h_{cr} in the greater measure has an effect on flow non-uniformity of various channels.

The expansion ratio is chosen 25, as is optimal from the earlier presented diagrams.

In this case size of the exit nozzle is 20 mm.

The coordinates of supersonic profile are determined on calculation method of nozzles for liquid rocket engines, the nozzle exit is performed at angle 4° between a center line of a blade and tangent to the profile of the nozzle. It is made for thin wall length reduction at nozzle exit. The obtained length of a supersonic part is 18 mm.

The subsonic part of a nozzle blade is made as of the low flow resistance envelope and has length of 25 mm (for increase of weight and rigidity of a blade). Critical cross-section has a straight line strip of 1 mm width.

Thus, having set the thickness of exit edge of 0.5 mm, we determine all remaining sizes of a profile of cross section of a nozzle blade (see Fig. 18.19b).

Beside, the selection of the thickness of exit edge in many respects depends on the material of a blade and technological capabilities of producer. Basically this thickness should come nearer to zero for elimination of axial wake behind each blade, however acute edge is technologically difficult to maintain along blade height edge, is inclined to destruction at any mechanical loads or oscillatory

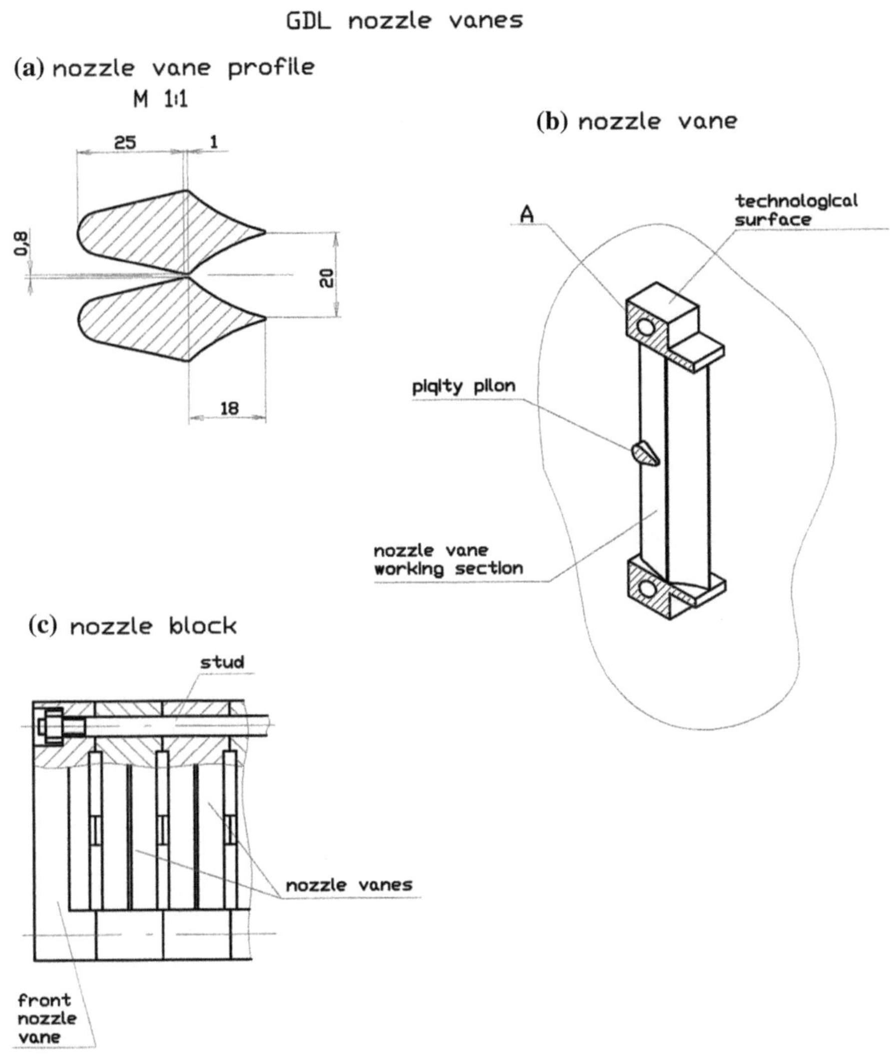

Fig. 18.19 GDL nozzle vanes **a** nozzle vane profile, **b** nozzle vane, **c** nozzle block

processes by operation. Therefore thickness of an edge is set in view of mechanical characteristics of the chosen material of a blade.

Nozzle blade as design unit (Fig. 18.19a) consists of a profile part two technological regiments and pylons for rigidity.

For classical of supersonic expansion nozzle form the critical cross-section of the nozzle should pass in a supersonic part through an acute edge. Thus all acute edges of critical cross-sections of blades should with high accuracy coincide with each other at the of blade assembly process in formation of package (Fig. 18.19a). For this purpose for assembly base orifices the precession orifice type in regiments of the blade is adopted. At blade set assembly with precision studs coincidence is provided of all decisive points of a profile of the neighboring blades. The size of critical cross-section h = 0.8 mm (between a critical point of a profile and regiment plane) is provided through final machining of all surfaces lying in a plane "A" (Fig. 18.19b).

The rigidity pylons exclude stretch of blades, and, hence, change of the size of a profile (especially on critical cross-section) in GDL operating time.

For tight leak connection of a package of blades in NU, surface "Á" of assembled with studs package is machined for obtaining high quality surface.

Quantity of blades in the package—81 pieces.

18.5.3.4 Supersonic Diffuser

The purpose of use is overexpanded supersonic flow pressure recovery to close to atmospheric value through transforming in subsonic flow.

The supersonic diffuser should have an inlet confusor (supersonic flow), critical cross-section (sound flow) and exit part (subsonic flow). In gas dynamics the technique of calculation of axisymmetric diffusers is state of art. For this case diffuser has rectangular cross-section 142 × 1,600 mm, therefore it is most rational for reduction maximum size to divide it in a number of channels with the help of internal contoured pylons.

As design (see Fig. 18.20) the diffuser represents a body of the box-shaped form, inside which the pylons are installed at a inlet angle 6° and exit angle 8° by screws.

18.5.3.5 Optical Resonator

The optical resonator as the structure is perimeter confined frame, mirrors and of the resonator adjustment gear. The design concept of the resonator is shown in Fig. 18.21.

The frame consists of two identical lateral metal racks. The rack has a window for transiting a laser beam, a place for attachment of adjustment gear and elastic lip. The elastic lip seals an interface of place of transiting of the rack through the chamber of the resonator and exclude transfer of vibration from working GDL to the optical resonator. The racks are connected among themselves by pipes of 200 mm diameter made from carbon-polymer materiel. It provides at sufficient rigidity, small weight, excludes temperature resonator deformation having practically to zero thermal expansion factor.

For elimination of acoustic resonance inducing effect on pipes they are filled by rubber sponge.

The mirrors of the resonator are made from high purity copper with protection of a working surface by of gold layer. A reflection coefficient of mirrors is 98 %.

One of mirrors (solid piece) has a concave spherical surface. On other, flat, side of a mirror there are places of attachment of adjustment gears through which

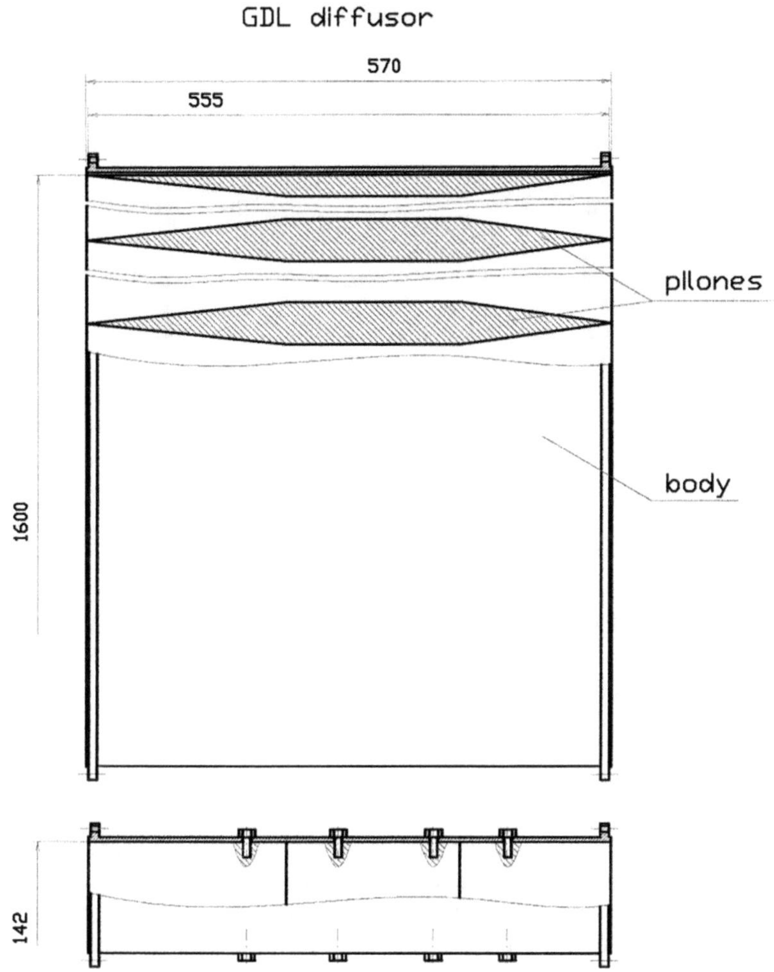

Fig. 18.20 GDL diffuser

the mirror is fasten to the rack of the resonator. Along the axis of the mirror the through hole $\varnothing = 1$ mm for resonator adjustment by the helium–neon laser.

The second mirror (translucent) has more complex form.

Directly mirror there is a convex spherical surface, around of it a ring slot is formed for transmission of laser radiation from the resonator. The elements of an adjustment and attachment of a mirror to the rack are placed on the external ring holding a convex mirror with the help of four flat thin pylons.

Convex mirror, the pylons and external ring are made as one piece.

As adjustment gears it is supposed at the first development phase to utilize the simple elastic holders with adjustable screws, which are used in laboratory lasers.

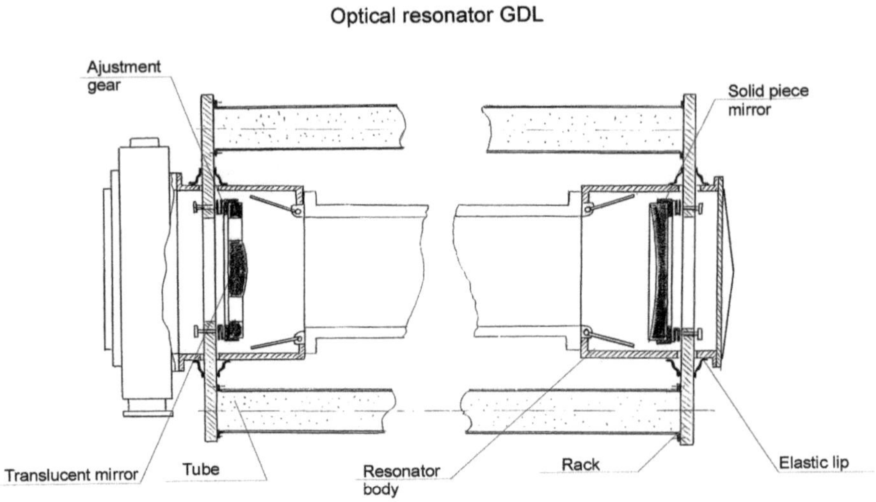

Fig. 18.21 Optical resonator of GDL

In future, if the laser installations will find broad application it is necessary to develop automatic adjustment gears. Some companies had conducted already activities in this direction.

18.5.3.6 Aerodynamic Window

The aerodynamic window serves for extraction of a laser beam of ring cross-section from the resonator in an environment. By laser operation at rather small altitudes atmospheric pressure is much higher than pressure of gas in the resonator, that can result to leak of free air in the chamber of the resonator and reduce the characteristics of the laser. The window from a transparent material cannot be utilized for the power laser, because of a partial absorption of radiation by the material of such window will cause it superheating and collapse. In the aerodynamic window the orifice for a transmission of radiation is overlapped by a supersonic jet. The pressure differential between ambient atmosphere and chamber of the resonator is organized through a system of shock waves along a bending of a supersonic jet.

The most economic schemes of aerodynamic windows, in which the working gas flow rate is minimum, are the windows realized through system of waves of compression and expansion with free swirl.

In presented case the cross section of a laser beam has the form of a ring with external diameter 125 mm. The inner diameter depends on a gain factor of an laser optical system. In the previous report [3] was shown, that the optimal gain factor is in range M = 1.5–2. Taking into account, that with increase of the area of the aperture of the window the flow rate of working gas of the aerodynamic window is increased proportionally of this area, and, that the requirements to an angular divergence are not too high in case of application of the laser for disposal of a

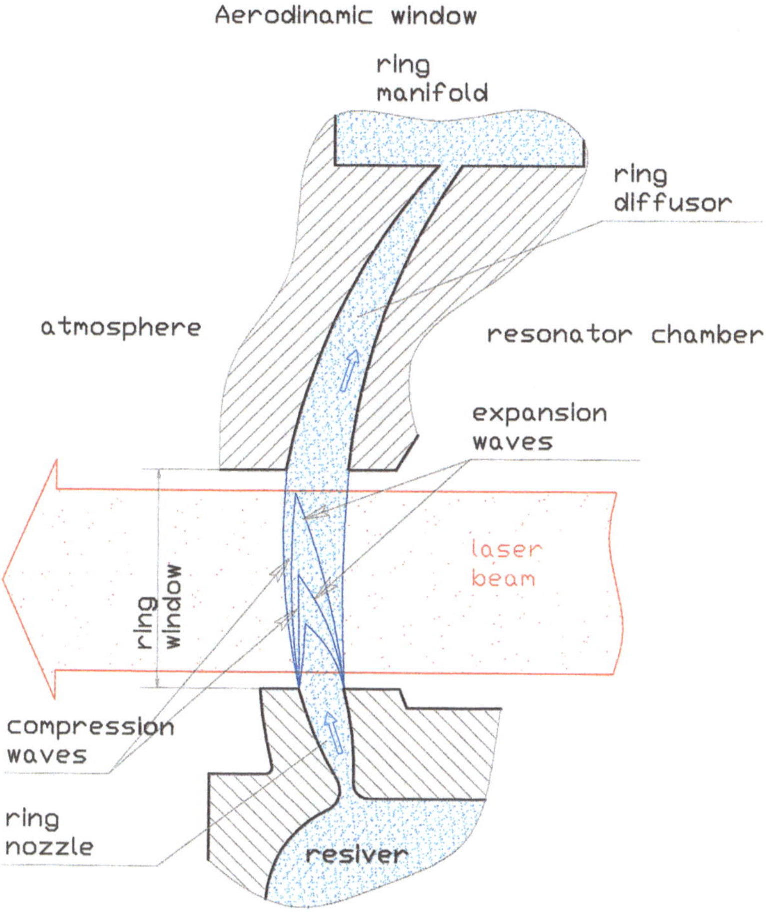

Fig. 18.22 Aerodynamic window

petroleum film, we shall accept M = 1.6. Then the inner diameter of beam cross section is equal 78 mm. To guarantee free transmission of a beam through aperture of the window it has a little bit more cross section than beam and represents a ring orifice with external diameter 130 mm and inner 73 mm.

The scheme of the aerodynamic window is shown in Fig. 18.22. Basic elements of the window—ring nozzle and ring diffuser. The ring nozzle is inclined in the side of window exit in a direction of a gradient of pressure. The ring diffuser is formed by two conic surfaces inclined in the opposite directions. Thus, the jet of working gas which is coming out of the nozzles, is directed to the part of high-pressure zone. Under an operation of a pressure differential the jet of gas deploys and enters in the ring diffuser. In the diffuser the supersonic flow of working gas expands, transforms in subsonic flow and then will be rejected in an exhaust GDL line.

The pressure across a jet of working gas of the aerodynamic window changes from atmospheric ~1 kg/cm^2 down to chamber pressure of the resonator ~0.12 kg/

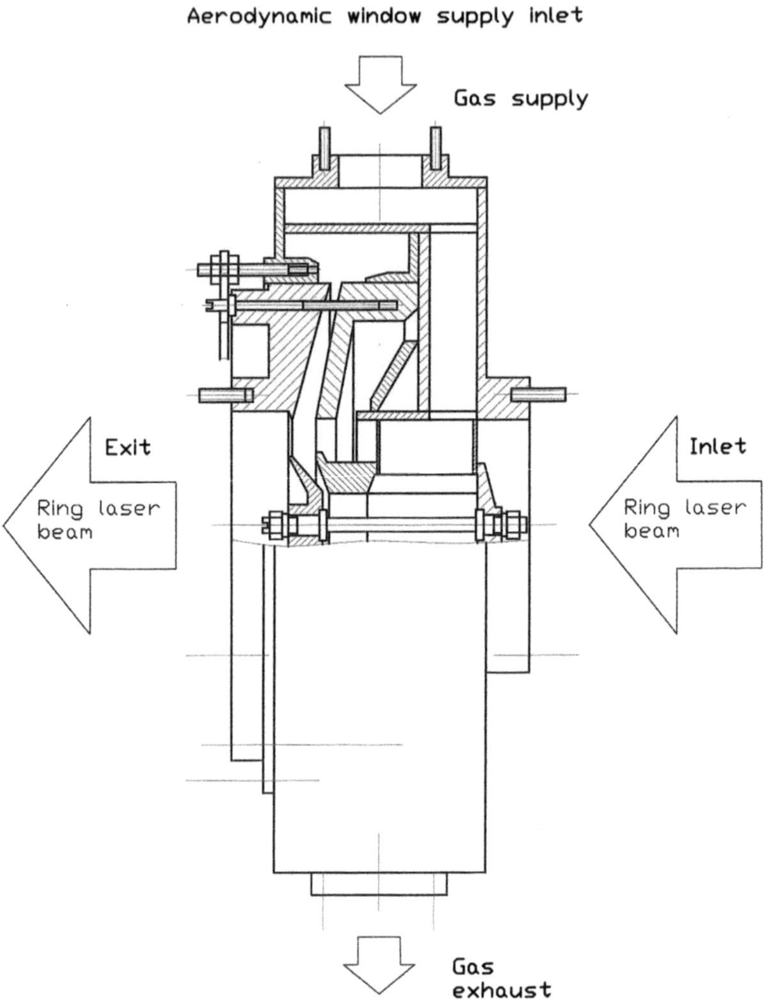

Fig. 18.23 Aerodynamic window supply inlet

cm^2. The change has taken place in waves of compression on boundary of a jet and ambient atmosphere, in gas jet in case of special contouring of the nozzle and in waves of expansion on boundary of a jet on the part of resonator chamber.

In this application ring nozzle of the aerodynamic window is countered so, that a Mach number in exit nozzle gas flow changes from ~3 at boundary of a jet with atmosphere up to 3.5 at boundary of resonator chamber. Therefore within jet the pressure changes approximately twofold. In waves of compression and expansion the pressure changes in ~5 times.

The scheme of flow of working gas through the aperture of the aerodynamic window are shown in Fig. 18.23.

In Fig. 18.23 it is possible to see the scheme of the aerodynamic window working gas supply. The gas in the receiver of the nozzle is suppled through flat pylons supporting the nozzle and a receiver. The pylons intersect a zone of propagation of laser beam. Their thickness and arrangement correspondent to thickness and arrangement of pylons supporting an output mirror of the resonator, that the shading of a beam was minimum. The collector ring established behind the diffuser of the aerodynamic window, has an exhaust branch pipe connected to a tube of gas removal down the diffuser.

The aerodynamic window prevents air leak in the resonator chamber at the atmospheric pressure to resonator pressure ratio ~10. Air will be utilized as working gas of the aerodynamic window with the flow rate ~2 kg/s. The feed pressure of working gas is ~20 kg/cm^2.

18.5.4 Exhaust Device of the Laser Installation

During gas-turbine engine operation the hot gas down the turbine exhausts with a high speed from the aft nozzle. At use of the engine on direct assigning (aircraft engine) the aft jet produces the additional thrust for the engine, is not limited to space and in a curtain degree is limited on acoustics.

During the use of the gas-turbine engine for the laser installation used on ground or on water vehicles of the requirement to a system of an exhaust of gas are much harder:

1. The acoustic oscillations of a jet should be minimized, as the installation can be applied on water reservoirs near to populating areas.
2. The exhaust should not create directional thrust vector (especially for the installations located on water vehicles), which can give all installation the non-authorized movement.
3. The exhaust jet of the engine should be mass of aft ejector working medium, ensuring operation of GDL supersonic diffuser.

Last problem is considered explicitly in the previous report, estimated performances and schematic of such ejector are added there.

To minimize acoustic oscillations (force of a sound reduction) during the exhaust from the engine, behind an ejector the exhaust silencer is to be installed.

There are three types of silencers:

Active reactive and combined.

The silencers of an active type are based on an absorption and dispersion of oscillatory energy at the expense of moving against the frictional forces of gas and such devices work on a pressure differential. For analyzed case they are not applicable since all excess pressure will utilize for ejection.

The silencers of a reactive type work on a principle of the acoustic filter blocking transition of oscillations of certain frequency (dangerous to human) in a system behind silencers.

The combined silencers combine principles of active and reactive damping and also require some jet overpressure for obtaining a pressure difference.

For the given project the reactive silencer is chosen. It is necessary to note, that in a design of such silencer it is possible to utilize a cascade principle and to provide adjustment of its elements on some characteristic frequencies, that provides smoothing of oscillations in more broad band of frequencies. Structurally scheme of such silencers is constructed on a combination of tubular resonators acting as of acoustic capacities.

One of varieties of such silencers is the device developed by "Pulseless control" company (USA) for damping compressor acoustic oscillations. The device of such type has the greater selectivity and allows considerably lower noise level.

The such silencer development consists of stringent selection, both supply lines, and silencer elements, for example cutting of acoustic tubes after the experimentally obtained values of frequencies and amplitudes caused by the given gas jet.

It is well-known, that such the acoustic filters can pass oscillations up to boundary frequencies

$$f = \frac{1}{\pi}\sqrt{\frac{1}{MaCa}};$$

where:

$Ma = \frac{\rho l}{F_t}$	acoustic mass of a tube;
$Ca = \frac{V}{\rho c^2}$	acoustic capacity of the chamber (volume V);
ρ	density of gas;
l	length of a tube;
Ft	he area of a tube;
c	rate of propagation of acoustic oscillations (velocity of a sound).

Thus, after obtaining the experimental data under the characteristics of acoustic oscillations at an exhaust of combustion products there is no difficulty to calculate and to design an acoustic silencer reducing noise level.

And the last requirement to the system of an exhaust of the laser installation—the exhaust gas thrust vector zeroing is determined by the customer.

Depending on place and method of arrangement of the laser installation the flow of exhaust gases can be divided, for example, on two opposite flow directions. It is possible to divide the exhaust around the circle to some weaker flows (this version is shown on applied figure) and at some stationary arrangement of the installation is permissible not to zero thrust vector, and to direct it in direction suited for the customer.

As explained above, the exhaust device of the laser installation can represent by design shown on Fig. 18.24.

All subsystem consists of three sequentially installed aggregates:

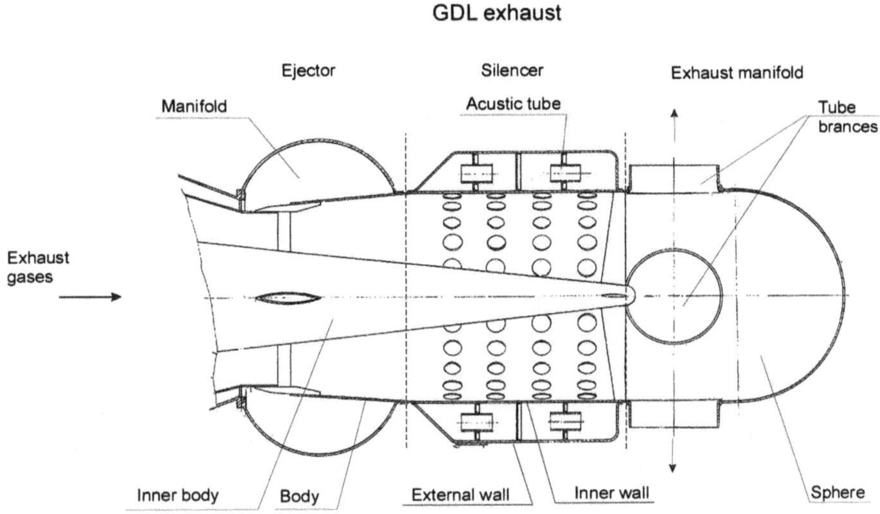

Fig. 18.24 GDL exhaust

Ejector, silencer and exhaust collector.

- The ejector consists of bodies with a collector of gas supply from GDL and inner cone. The walls of a body and collector form cone—inverse cone—calculated profile, in minimum cross section region of which—is organized supply after GDL.
- The silencer represents the multisectional acoustic chamber, has outside and internal walls. The internal wall has a number of orifices connecting the chamber with a gas flow. The sections are divided by walls and in each section the acoustic tubes, selected on the sizes according to the acoustic characteristics of an exhaust system (acoustic adjustment of a silencer) are established.
- The exhaust collector represents spherical container with a number of opposite directed branch pipes, through which the gas expires.

As it was indicated above, this part of an exhaust system is added as version, since its substantial design is depending on place and operation conditions of the laser installation.

18.5.5 Other Systems and Aggregates of the Laser Installation

The laser installation includes:

- Directly GDL—autonomous installation (is described above); with the systems
- Engine HK-29—power unit ensuring GDL by air and the electric power (was considered in the previous report).

- Exhaust device playing the role of a gas ejector, reducing installation noise level and zeroing (if necessary) an exhaust jet thrust of the installation. In the report its design and description are added.
  Besides, in installation enter the systems and aggregates, which design in the report is not considered, as they do not represent technological problems and can be in most cases borrowed from serially made design assemblies and products of industry.
  The following are included:
- Container for kerosene storage. It is a tank with a volume ensuring a required operating time of the installation.
- Fuel and power supply system and gas tubes. The pipes, adapters, flanges etc. can be selected from used in aerospace industry. For compensation of temperature deformation and the eliminations of transmission of vibrations lines connecting GDL and aviation engine, are connected through bellows or flexible sleeves.
- External optical system. By development of power lasers, which radiation was transmitted to significant distances (hundred and thousand meters) usually the external beam forming optical system with the purpose of beam divergence reduction is used.

In presented installation and at use GDL for clearing of water surface, distances between GDL and surface of a water is no more than several meters. Therefore employment of complex and expensive telescopic external system is not profitable. It is enough to place on a course of a laser beam a flat rotary mirror, with which it is possible to scan a beam on a surface of water.

The control panel. The engine HK-29 is supplied with the control panel for engine start and shutoff. For GDL operation control it is necessary to finish the existing console by introduction the additional printer circuits on formation of commands on GDL control elements and monitoring display. Apparently, it will be by a workplace for the second operator.

18.5.6 GDL Installation

In view of all computational investigations, design studies and engineering estimations within the present report works the provisional structure of the autonomous laser installation for clearing of surfaces of reservoirs of petroleum recordings is produced.

The general view of such installation is shown in Fig. 18.25.

As it is visible from figure, the installation represents modular assembly.

The engine has an own fixing frame, autonomous subsystems and producer—supplier.

GDL also represents the complete aggregate, the developer and manufacturer of which is not connected to other suppliers.

The same concerns to fuel storage system (is not shown in Fig. 18.25).

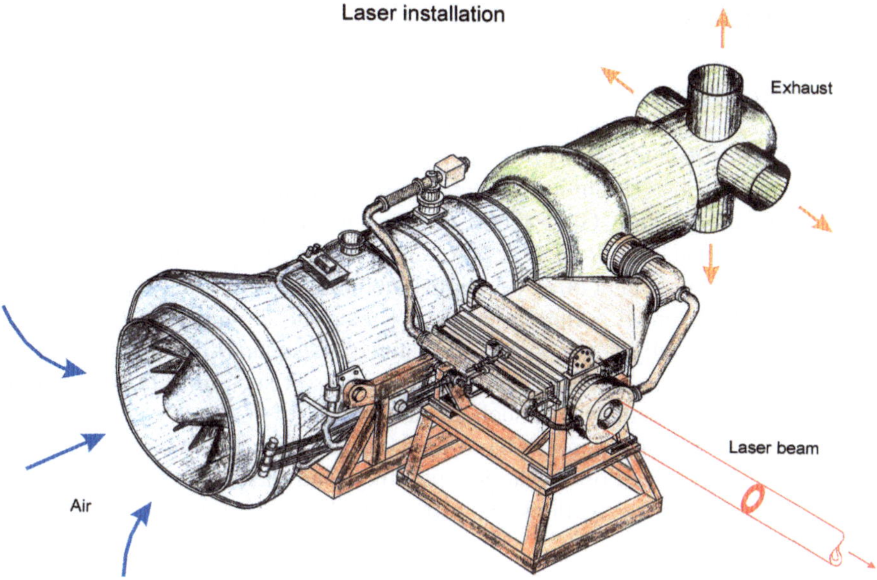

Fig. 18.25 Laser installation

Task of the installation developer as a whole is the coordination of these units both on arrangement on a vehicle, and organization of interaction during operation.

For example, relative arrangement of the engine and laser can be arbitrarily, depending on overall dimensions and device of a vehicle, however in any case the fulfillment of following conditions is necessary:

- The engine and GDL should not have of rigid connection on a frame, the GDL frame should have vibration absorption plates.
- The engine and GDL should be placed as close as possible, as the unreasonable elongation of gas lines connecting them, will result into additional losses because of pressure reduction.
- On fire protection control requirement fuel storage container should be placed on distance from the engine and GDL, and is separated from them by a fire-prevention wall.
- For protection of the attendants the remote control panel should be born for limits of the installation, and input part of the engine and line of a beam are protected by special fences. The control of a scanning mirror also is carried out for limits of the installation.
- The direction of exhaust gases should not intersect structural items of the installation.

As the development of the laser installation does not assume development of testing site for development testing, prior to deliveries the development installation should be assembled, on which all dezigner and operational problem solutions will be find.

18.6 Conclusion

The Purpose of presented project is to develop a laser technology of cleaning of large areas of seas and other water surfaces from oil film contamination. This paper also aims development of composition scheme, technological requirements and engineering design solution of aggregates and units of vehicle, vessel or aircraft compatible mobile GDL (100–250 kW) intended to solve this important problem of environment protection [4–6]. This method proposed for development is expected to complement other traditional methods, which usually more successfully treat bulk layer oil pollution but do not match to eliminate up to 100 % oil films, the latter usually being spread over numerous square kilometers of sea and kilometers of seacoasts. Note about one million tons of such films drift now in the World Ocean, as consequence of such disasters.

Thus, this promising approach has a fundamental science basis, from one hand, and skill and experience of missile jet specialists and their laser prototypes, from another hand. The investigations are planned to be accomplished in three basic directions:

(1) theoretical and experimental research of the laser radiation action on thin oil and it's derivatives films over water surface, evaporation in three basic directions:
(2) experimental optimization of laser optical scheme and operation modes in order to achieve the maximum optical and gas—dynamic efficiency.
(3) engineering design, composition scheme and accompanied problems solution in order to adapt the laser module for vehicle, ship or aircraft transportation.

Results expected are as following:

- Certain recommendations on oil films elimination from large water areas, including conditions of real tanker and pipeline disaster, by means of laser radiation; development of technologies of oil and it's derivatives films burning out (mode number one) and oil films sucking up and saving of oil products for future efficient usage (mode number two) by means of high power mobile CO_2 GDL.
- Development of self-contained mobile CO_2 GDL tailored for these environmental problems solution.

These investigations may result in significant progress in the following branches of science:

(a) physics of liquid inhomogeneous films, phenomena on the boundary of water-carbohydrate composition, phase transitions and chemical processes in two-phase or multi-phase compositions under high power CO_2 laser radiation and related phenomena;
(b) problem of active resonators for high power GDL, different methods of laser efficiency calculation and simulation, control of temporal radiation modulation and optimization of performance, with respect to the problem of beam quality.

The author has completed the conceptual analysis of GDL installation development for disposal of petroleum films from a water surface. The basic capability of development of such installation is shown. The basic characteristic calculations of the system are made, the schematic and design solutions of basic installation are presented.

Besides of that, the GDL bench has been restituted by the author of the paper and his colleagues on their own costs and some demonstration operations with scanning the beam over water surface covered with petroleum film were carried out. The experimental confirmation of film effective burning mode and mode of oil film gathering was obtained. The next part of realization phase can be the development of the installation preliminary design, in case of the customer objective and specific technical tasks, formulated by him. The preliminary design will include complete study of a design of all elements of the installation, technological works, will define all suppliers of the developing systems and sub suppliers from obtaining from them of the consent on terms and conditions. In the preliminary design, schedules of all phases of installation development will be worked out: issue of the working documentation, technology modification, manufacturing of a prototype, prototype development and beginning of installations deliveries.

Thus, according to the estimation the experimental mobile CO_2 GDL installation for water surface cleaning can be designed, manufactured and tested for not longer than two to two and half years.

References

1. V.V. Apollonov, A.M. Prokhorov, Universal laser for industrial, scientific and ecological use, in *Proceedings of GCL/HPL Conference*, St. Petersburg, 1998, p. 140
2. V.V. Apollonov, Ecologically safe high power lasers, in *Proceedings of Lasers-2001 Conference*, 2001, p. 3
3. V.V. Apollonov, High power autonomous CO_2 GDL (100 kW, CW/P-P modes) for new technologies development and environment protection; Russia –NATO International Seminar, Moscow, 2005, in *Proceedings of GCL/HPL Conference*, Lisbon, 2008, p. 35
4. V.V. Apollonov, New application for high power high repetition rate pulse-periodic lasers, in *Laser pulse Phenomena and Applications*, Intech, 2010, p. 19
5. V.V. Apollonov, Oil films elimination by laser, Oboronzakaz, №17, December (2007), p. 33
6. V.V. Apollonov, Yu. S. Vagin, V.V. Kijko, High repetition rate P–P lasers, Patent RF № 2175159

Chapter 19
High Power Lasers as a Weapon

19.1 Introduction

The experience of designing strategic laser systems in the previous years and a strong belief in the attainability of the goal—the development of high-power laser weapons—help to significantly accelerate the pace of work in the field of new technologies. World press has recently published the articles where it was suggested that the anti-missile airborne laser project suffers cost overruns and delays and may fall victim to budget cuts. Just after that mass media immediately reported that the Pentagon refuses to develop combat lasers. As a result, there has appeared an epic set of myths about combat lasers. The task of this part of the book is to consider the most ridiculous statements and to reject them by sufficient analysis.

Today, chemical lasers are being widely replaced by solid-state diode-pumped lasers. It is these lasers that the Pentagon counts on, because they are more compact, simpler and cheaper to use than chemical lasers. Besides, they are reliable, easily compatible (without any transformation of the output voltage) with nuclear and solar energy, allow for further scaling of output parameters and the efficiency of their operations is significantly higher. Northrop Grumman Corporation has already presented a 105-kW solid-state laser and intends to significantly increase its output power [1]. According to data from U.S. laboratories work is on the way to develop a prototype of a 500-kW laser. Subsequently, 'hyperboloids' will be truck-mounted [high energy laser technology demonstrator (HEL TD) program], ship-mounted [maritime laser demonstrator (MLD) project], and airborne [high energy liquid laser air defense system (HELLADS); a laser for F-35, B-1, CH-47 aircrafts). Another direction is largely supported by Raytheon, which stakes on fiber systems. A 50-kW laser weapon system (LaWS) will be integrated with the Phalanx Close-In Weapons System (CIWS) against anti-ship missiles and its land based version of Centurion C-RAM. In addition, the officials in the U.S. have recently reported on the progress of works on a combat free electron laser. At the same time, we should not forget about "Alpha" system (a laser with an output power of 4.5 MW), lying on the ground and waiting for the decision to be launched.

V.V. Apollonov, *High-Power Optics*, Springer Series in Optical Sciences 192,
DOI 10.1007/978-3-319-10753-0_19

Anti-missile airborne laser

Brining solid-state lasers to megawatt powers takes time and considerable resources. However, the experience of designing strategic laser systems in the previous years and a strong belief in the attainability of the goal—the development of high-power laser weapons—help to significantly accelerate the pace of work in the field of new technologies. It should be noted, however, that tactical laser systems with lower power outputs have already been tested in the U.S. and now find applications in the army. All this testifies that the Pentagon experts clearly do not think about cancelling the promising laser programs. We are talking here about an effective system of disinformation. Last year's report 'Technology Horizons' by the Pentagon refers to the global changes in the 'rules of the game' after the proliferation of 'high energy weapons', which will turn the traditional symbols of military power into outdated stuff like cannonballs and cavalry.... And while the U.S. develops laser programs, many other counties show 'laser apathy.' Incompetent bloggers and pseudoscientific workers, who had something to do with the laser program 30 years ago, speculate that 'lasers are a bluff.' As a result, there has appeared an epic set of myths about combat lasers. Consider the most ridiculous of them.

19.1.1 Myth 1. "Combat Lasers have been Developed for Four Decades with no Progress in Sight."

Let us quote one of the Russian papers: "In the 1970s the Americans took a 150-ton Boeing-707 and 'stuck' a laser to it, which successfully burned small rockets. In the 2000s, they took a 350-ton Boeing-747 and 'stuck' a heavier and more powerful laser, which successfully burned larger size rockets. In 20 years they will buy in Ukraine a decommissioned "Antonov AN-225 Mriya" (640 tons), and here it is, the Death Star [2]. Yes, probably it will be able to shoot down a "Scud", and even a "Taepodong". However, only at the site, and once, no more."

Under the '150-ton Boeing-707,' to which a laser was 'stuck,' is apparently meant a 137-ton Boeing KC-135 stratotanker based on the "707" of the first), modified in 1973 into NKC-135ALL [airborne laser laboratory (ALL)]. In 1983, a laser mounted on the aircraft shot down several Sidewinder air-to-air missiles at a distance of 5 km and some other 'little' things. What has changed since then? According to the above quotation—only the size of the aircraft.

Airbased military laser

But what about the reality? Even the so-called 'megawatt' CW lasers of the 1980s did not emit light at megawatt power levels, they consumed more power. The 2.2-MW Miracle laser system, which later in the combat version was called a tactical high energy laser (THEL) ('MIRACLE' with the 'SEALITE' guidance system) did not possess a supernatural power [3]. What to say about the earlier and five times weaker ALL. Has there been any progress since then? An ABL has a power of 1.1 MW and it is not the power to be consumed; this is the power in the beam. Thus, a 'more powerful' (50 times) laser was 'stuck' to a 350-ton Boeing.... However, it should be understood that the actual capabilities of the laser rely not on the power as such, but on the concentration of the radiation, i.e., the ability of the 'guns' to emit not only a powerful, but also a narrow beam. The ALL had the level of the radiation concentration equal to 10^{13} J/(sr·s). An ABL has the level of about 10^{18} J/(sr·s)—that is, 10 thousand times higher. These achievements are made up not only of the straight-line growth in the power. The last 30 years have seen a period of extremely rapid development of adaptive optics to compensate for the effects of atmospheric turbulence and laser path on the transmitted beam. In addition, the lasers of the same class are radically reduced in size. The first version of a THEL weighed 180 tons and could be hardly installed into six trailers. The laser used a hydrogen–fluoride mixture which is very unfriendly to the environment. The second generation of advanced tactical lasers (ATLs) relies on oxygen–iodine mixtures [so called chemical oxygen–iodine lasers (COILs)] and is more compact. Finally, a new solid-state laser of Northrop Grumman weighs 1.5 tons, including the cooling system. In the future its weight is expected to be reduced to 750 kg. As a result, the land based version of the system consists of a heavy expanded mobility tactical truck (HEMTT A3), command post on a high mobility multipurpose wheeled vehicle and a towed single-axle trailer with the AN/MPQ-64 radar. At the same time, the U.S. works hard to convert the CW regime of the lasers in the pulse-periodic one, which will dramatically increase their range of actions.

Mobile laser system

Talks about the fact that "combat lasers have been developed for 40 years and so they are hopeless," only show ignorance in technical matters. Breakthrough technologies are first tested for a few decades before the entry into a phase of maturity. Thus, aircrafts at the time of the first flight had almost 60 years of history—the first flying models were built in 1840 and the attempts to build full-size airplanes date back to 1868. It is, in fact, a classical scheme of development of any technology using new physical principles: First, a long 'incubation period' with no apparent practical results, and only then a 'great leap forward.'

19.1.2 Myth 2. "Lasers Cannot be Used for a Long Time, Usually they Work Several Seconds."

It is not so! In fact, chemical and solid-state combat lasers ensure continuous wave operation for minutes and tens of minutes. The next step in the development of high-power laser systems, of course, will be the implementation of a variable temporal structure of radiation in order to raise the peak radiation power for the ablation mechanism and to eliminate the effect of the screening of the target by plasma [4].

19.1.3 Myth 3. "Energy of Laser Weapons is Negligible Compared to the Fire Arms."

For comparison, the power of a 76.2-mm F-22 divisional gun (1936) is at 150 MW. This power is 150 times higher than that of the ABL! Besides, we ignore here the energy of the explosive in the shell, which is the same. Ponder upon this simple fact:

A small ancient gun of the Second World for the price of scrap metal is hundreds of times more powerful than the ultramodern 'combat' laser weighing tens of tons and valued at over five billion USD. Only one shot from the ABL costs millions of dollars. And this shot comparable in energy with the burst of a machine-gun fire.

Such a comparison of the power achieved in 0.01 s with the power of CW light, and the 'proof' of inferiority of 'long-play' weapons (by using this comparison) contradicts even the school physics course. Let us try to compare everything in the correct way, i.e., by counting the energy sent to the target.

The muzzle energy of a 12.7-mm heavy machine gun is 15–17.5 kJ at a rate of 80–100 shots/min. In other words, a 100-kW laser can be 'replaced' by three and a half heavy machine guns (6,000 kJ/min vs. 1,750). But let us return to the gun. The muzzle energy of F-22 is 1.35 MJ, while the power of the ABL is 1.1 MW, i.e., 1.1 MJ every second. Thus, the laser shoots 48 'shells'/min. By converting the MW power to the TNT equivalent, we will obtain 240 g of explosives per second and 14.4 kg/min, which corresponds to the content of 18 high-explosive shells from the same gun. However, the actual 'value' of the laser is higher. The matter is that even with accurate firing of firearms, most of the 'energy' goes not to the enemy, but dissipates. The reason for that includes a dozen factors (wind, fluctuations in humidity, air pressure and temperature, the Coriolis force, etc.), making the bullet or shell spread inevitable. A steady flux of photons flies to where it was sent, excluding the huge amount of unnecessary loss.

19.1.4 Myth 4. "Efficiency of Lasers is a Few Percent."

In fact, the efficiency of combat lasers is about 20.6 %, and this is not the limit. According to the RELI program, the efficiency is planned to be raised up to 25 %. Fiber lasers, which are adapted by Raytheon for military applications, now have an efficiency of about 30 %. The efficiency of firearms is 20–40 %. At the same time, the efficiency of (smaller but steadily growing in power) solid-state diode-pumped systems is more than 50 %, and soon it will approach its physical limit of about 85 %.

19.1.5 Myth 5. "The Laser Beam has a Huge Diffraction Divergence."

Here we deal a physically insuperable law of diffraction, which states that the laser beam always diverges with an angle proportional to the ratio of the wavelength to the diameter of the beam. If we take a specific combat IR laser with a wavelength of 2 μm (this wavelength is typical of combat THELs, etc.) and a beam diameter of 1 cm, we obtain a divergence angle of 0.2 mrad (this is a small angular divergence; for example, standard laser pointers/rangefinders have a divergence angle of 5 mrad and higher). However, because of 0.2-mrad divergence the beam spot diameter will increase from 1 to 3 cm at a distance of 100 m. That is, only 100 m away the beam density will decrease (proportionally to the area of interaction) by seven times. At a distance of a kilometer the density of the beam falls by 300 times.

Actually, a combat laser emitting a beam with the initial diameter of 1 cm is a fruit of an unhealthy imagination, not burdened with some knowledge in this area. In fact, when using the focusing optics the diffraction divergence is about λ/D, where λ is the wavelength, and D is the diameter of the mirror and also the initial beam diameter, tapering due to focusing as the beam approaches the target; a large beam diameter (meters in this case) provides a low diffraction divergence. In the case of the ABL the wavelength is 1.315 μm, and the diameter of the mirror is 1.5 m; by dividing one by the other, we obtain the divergence of about 10^{-6} rad. In other words, at distance of only 1 km the laser beam from the Boeing diverges by (oops!) 1 mm. At a distance of 200 km, the diffraction divergence will be 20 cm. The actual ABL beam divergence exceeds the diffraction limit by only 1.2 times.

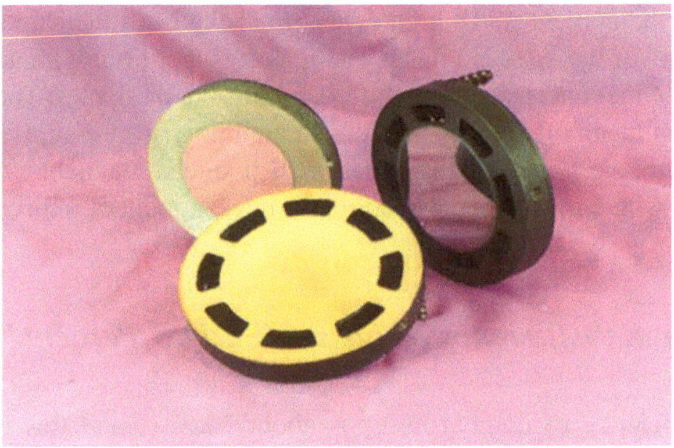

19.1.6 *Myth 6. "One can Easily be Protected from Laser Weapons Using, for Example, an Aluminum Mirror."*

Yet another malapropism. Indeed, some metals may have near-100 % reflection coefficients. First, however, these coefficients are not equal 100 %. Thus, at a wavelength of 1 μm the reflection coefficient drops to 75 % for most structural metals. A real missile after its launch will be also significantly contaminated with combustion products. Meanwhile, modern 'hyperboloids' emit light in the vicinity of 1 μm (the ABL has a wavelength of 1.315 μm). In this case, 25 % of hundreds of kilowatts will be sufficient even in the cw regime to heat up and melt down the thin skin layer of the missile. Thus, reflectivity will decrease because the absorption of laser radiation increases rapidly with increasing temperature, and abruptly jumps after the start of melting. In the pulse-periodic regime the situation is even more favorable.

Besides, there arises also a 'childish' question: If the laser beam can be focused and directed by a mirror, why cannot the same mirror protect us from the laser beam? In lasers use is usually made of multilayer dielectric mirrors that can reflect very much—but in a very narrow range and only under strictly defined angles [5]. In addition, they are cooled, which is impossible to do with the entire surface of the target. In other words, simple, effective and affordable protection from high-power lasers does not exist.

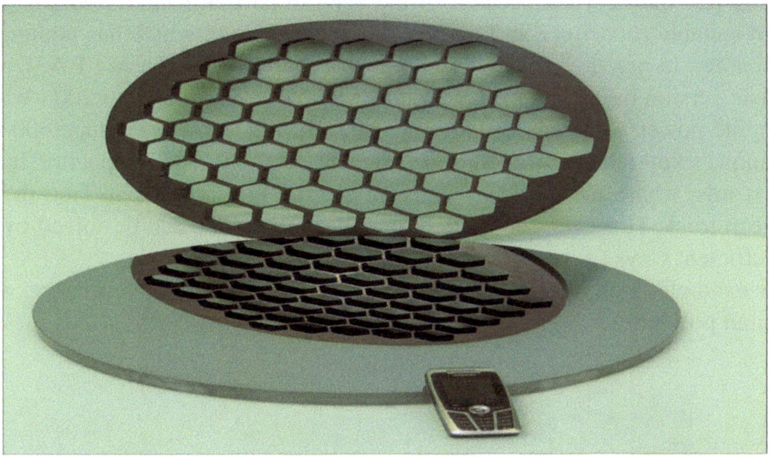

SiC mirror and SiC frame for cooled mirror

19.1.7 Myth 7. "The Problem of Overheating of Lasers Cannot be Solved."

Four megawatts of heat, which can heat red-hot an aircraft and burn to the ground, are generated per each megawatt of laser power. The cooling system with the gas flow rate of 1,800 m/s (de Laval nozzle) is unable to release the generated heat from the fuselage.

In reality, the 'disposal' of the amount of heat in the megawatt units is quite trivial. Has anyone seen a red-hot diesel locomotive? Meanwhile, a decent diesel engine with a capacity of 2 MW releases more than 1 MW of heat into in the oil and cooling system. Far less easy is the problem of heat release from the limited volume of the laser weapon. In the case of a chemical ABL, the heated reaction products are simply blown out of the cavity by the well-known de Laval nozzle, and then liquid ammonia is used for cooling. Fairly cumbersome is the system with cryogenic components; however, it really is able to 'recycle' a very impressive amount of heat. Tactical solid-state lasers, which need more than 400 kW of heat to be released, do well without cryogenic 'refrigerators.' Thus, the HELLADS makes use of a unique cooling technique; the circulation of a liquid releases excess heat outside the 'laser cannon.' Remarkable also is General Atomics' advanced thermal energy storage device capable of cooling directed energy weapons. Heat is stored in the 35 kg module by melting a wax-type phase change material. As a result, the HELLADS provides for the interception of missiles within 2 min at the specified cw power followed a 30-s break.

19.1.8 Myth 8. "High-Power and Compact Energy Sources for Combat Lasers do not Exist."

This is partially true: It is not yet possible to mount a 100-kW solid-state laser on anything less than a truck because of the need to have at hand a 500-kW generator and suitable capacitors. In fact, this is a real problem, which has nothing to do with fantasy. In practice, the hybrid version of the HEMTT (HEMTT A3), even in the basic version has a 350-kW generator, which can provide up to 200 kW. When the engine power is increased up to 505 hp, HEMTT A3 can produce 400 kW of continuous exportable power. A nice addition is the 1.5-MJ capacitor bank. In other words, where the bloggers fancy an electric power plant, in fact, there is a high-tech truck. However, the issue of energy in the space can be solved by other, more efficient ways.

For example, well developed are nuclear power sources, solar energy, with its unlimited possibilities.

Mobile laser system

19.1.9 Myth 9. "Every Shot a Laser is Worth Millions of Dollars."

In fact, one shot of the ABL costs 10 thousand USD, whereas the domestic 16 million rubles are the propaganda and exaggeration. This is comparable with the cost of a Fagot anti-tank guided-missile system. More complex anti-tank guided-missiles cost tens of thousands of dollars; for example, a Maverick air-to-ground missile for hitting targets ranging from a distance of a few thousand feet to 13 nautical miles at medium altitude costs 154 thousand USD, a Patriot missile—3.8 million

USD. The cost of a shot from tactical lasers is less than from the ABL. Even in a hydrogen fluoride THEL it was 2–3 thousand USD; in this case, the laser used not hydrogen but deuterium which is quite expensive.

Sketch space based laser system

19.1.10 Myth 10. "All the Problems that can be Solved with Laser Weapons, are Easier and Cheaper to Solve by Traditional Means."

This conclusion has already been proven to be ineffective; see, for example, at Israel's attempts to defend against missile attacks by Hamas missiles (the Iron Dome system). One interception costs 30–40 thousand USD. The cost of a rocket for a Grad multiple rocket launcher system is about 1 thousand USD, the cost of a Qassams rocket does not exceed 200 USD. Thus, the interception will be 40–200 times more expensive than the rocket launched by the enemy. In this connection Tariq abu Nazar, the Hamas spokesman, once said "if every rocket launch of our missilemen costs tens of thousands of dollars for Israelis, we will assume that the goal is achieved." As a result, some newspaper people accuse not developers of laser systems but those who are responsible for termination of the Israeli-American program. The Centurion missile defense system has not found wide application because of the short-range and great ammunition consumption.

19.2 Conclusion

Of course, this is not a complete list of the legends about lasers. Most of them are built on the same principle: either outrageous lie or painstaking making of a mountain out of a molehill. In fact, lasers on the battlefield are real and the army, which possesses such weapons, will receive an impressive advantage. Thus, the air force, able to actively defend against anti-aircraft and air-to-air missiles, will be much less vulnerable to air defenses. Thus, the development of laser technology is critically important not only for well developed countries. Combat lasers are an obvious asymmetrical response to the superiority of the precision weapons. The 'ideology' of the latter in a very rude manner is to ensure that instead of a dozen of shells, a guided munition (though much more expensive) is used to precisely hit a specific target, and to minimize collateral damage. However, such a scheme is particularly vulnerable to laser defense systems, which make no difference between an archaic shell for two hundred dollars or an expensive high-tech device or military equipment. The number of targets is not so great, and their cost is ten times higher than that of the most expensive laser 'shot.'

It is already evident that the world has entered a new round of technological race. It does not depend on our wishes, it is the willing of time. Most developed countries, based on their technological advantages, are spending billions of dollars to develop high-tech next-generation lasers [6]. According to Japanese media U.S. invested up to now more than 100 billion USD in the development of semiconductor laser pumping of solid-state lasers. The need to accelerate the development of high technology is the major topic of many political leader speeches. It is important to note the opinions of Western experts, who say that today laser technology is one of the most effective ways to gain technological superiority in the world. And today, laser technology makes it possible:

- to provide a new level of development of industry, science and technology for your country, the revival of scientific and technological strength of the country on the basis of modern high-techs;
- to make your country a leader in the field of technical, scientific and technological progress;
- to revive a large number of enterprises in many sectors of industry, which are well known for their developments in previous years;
- to strengthen your country leading position in space and to ensure strategic and geopolitical priorities in the modern world; and
- to derive benefit from the sale of a wide range of laser programs, which is today comparable to the benefit from the sale of traditional products and resources.

References

1. SPIE Events Europe, Nothrop's R@D, INTERNET (2012)
2. P.V. Zarubin, QE **32**(12) (2002)
3. J. Aerosp. Defense Ind. News (1997)
4. V.V. Apollonov et al., QE **33**(9) (2003)
5. V.V. Apollonov, Laser Phys. **23** (2013)
6. V.V. Putin, Message to the Federal Assembly of RF (2013)

Appendix

Patents and Certificates of Recognition

1. Certificate of recognition №103162 from 24 of May 1976
2. Certificate of recognition №114243 from 16 of May 1977
3. Certificate of recognition №122743 from 15 of March 1978
4. Certificate of recognition №123604 from 13 of February 1978
5. Certificate of recognition №123791 from 09 of February 1978
6. Certificate of recognition №126323 from 2 of June 1978
7. Certificate of recognition №135237 from 8 of December 1978
8. Certificate of recognition №135109 from 27 of November 1978
9. Certificate of recognition №137270 from 01 of September 1978
10. Certificate of recognition №139755 from 03 of April 1979
11. Certificate of recognition №144371 from 19 of January 1979
12. Certificate of recognition №142696 from 29 of June 1979
13. Certificate of recognition №142697 from 29 of June 1979
14. Certificate of recognition №753248 from 23 of February 1979
15. Certificate of recognition №144370 from 19 of January 1979
16. Certificate of recognition №153300 from 11 of February 1980
17. Certificate of recognition №157039 from 29 of May 1980
18. Certificate of recognition №152944 from 14 of February 1980
19. Certificate of recognition №989299 from 22 of May 1981
20. Certificate of recognition №1026594 from 27 of November 1981
21. Certificate of recognition №1017901 from 27 of July 1981
22. Certificate of recognition №180957 from 25 of February 1982
23. Certificate of recognition №180959 from 25 of February 1982
24. Certificate of recognition №184038 from 19 of April 1982
25. Certificate of recognition №190993 from 01 of April 1982
26. Certificate of recognition №190994 from 01 of April 1982
27. Certificate of recognition №1053696 from 26 of February 1982
28. Certificate of recognition №1037816 from 22 of February 1982
29. Certificate of recognition №1046984 from 14 of May 1982

V.V. Apollonov, *High-Power Optics*, Springer Series in Optical Sciences 192,
DOI 10.1007/978-3-319-10753-0

30. Certificate of recognition №1190895 from 06 of July 1982
31. Certificate of recognition №1108990 from 07 of February 1983
32. Certificate of recognition №202719 from 18 of July 1983
33. Certificate of recognition №1168011 from 20 of September 1983
34. Certificate of recognition №208958 from 05 of July 1983
35. Certificate of recognition №208289 from 19 of December 1983
36. Certificate of recognition №225424 from 20 of December 1984
37. Certificate of recognition №1268904 from 10 of November 1984
38. Certificate of recognition №225192 from 30 of November 1984
39. Certificate of recognition №1398718 from 11 of March 1985
40. Certificate of recognition №1398717 from 04 of April 1985
41. Certificate of recognition №1301260 from 08 of August 1985
42. Certificate of recognition №240743 from 20 of May 1985
43. Certificate of recognition №1398717 from 04 of April 1985
44. Certificate of recognition №1398719 from 11 of March 1985
45. Certificate of recognition №1294249 from 28 of January 1985
46. Certificate of recognition №1306441 from 31 of May 1985
47. Certificate of recognition №244286 from 10 of December 1985
48. Certificate of recognition №237574 from 26 of July 1985
49. Certificate of recognition №1278565 from 04 of April 1985
50. Certificate of recognition №1277854 from 28 of February 1985
51. Certificate of recognition №234217 from 27 of May 1985
52. Certificate of recognition №234216 from 27 of May 1985
53. Certificate of recognition №241100 from 02 of October 1985
54. Certificate of recognition №251080 from 14 of May 1986
55. Certificate of recognition №1389627 from 20 of June 1986
56. Certificate of recognition №1396896 from 14 of July 1986
57. Certificate of recognition №1336885 from 31 of March 1986
58. Certificate of recognition №261206 from 27 of October 1986
59. Certificate of recognition №1455217 from 24 of February 1987
60. Certificate of recognition №1514205 from 06 of October 1987
61. Certificate of recognition №1459570 from 17 of March 1987
62. Certificate of recognition №1435116 from 09 of February 1987
63. Certificate of recognition №1499688 from 09 of November 1987
64. Certificate of recognition №1455959 from 29 of June 1987
65. Certificate of recognition №288490 from 25 of December 1987
66. Certificate of recognition №1459570 from 17 of March 1987
67. Certificate of recognition №1592711 from 19 of March 1987
68. Certificate of recognition №289258 from 29 of April 1988
69. Certificate of recognition №1510560 from 10 of February 1988
70. Certificate of recognition №333228 from 05 of February 1988
71. Certificate of recognition №1586479 from 03 of May 1988
72. Certificate of recognition №234217 from 27 of May 1988
73. Certificate of recognition №1515979 from 04 of January 1988
74. Certificate of recognition №1672896 from 08 of June 1989

75. Certificate of recognition №1672897 from 08 of June 1989
76. Certificate of recognition №1753443 from 25 of September 1989
77. Certificate of recognition №322187 from 27 of September 1989
78. Certificate of recognition №330267 from 29 of November 1990
79. Patent №2017313 from 10 of May 1990
80. Patent №2047880 from 28 of June 1991
81. Patent №2117371 from 30 of September 1996
82. Patent №2169977 from 16 of August 1999
83. Patent №2130221 from 23 of April 1999
84. Patent №2166820 from 22 of January 1999
85. Patent №2166821 from 22 of January 1999
86. Patent №2166823 from 22 of January 1999
87. Patent №2166822 from 22 of January 1999
88. Patent №2181225 from 05 of June 2000
89. Patent №2176120 from 16 of May 2000
90. Patent №2175159 from 28 of April 2000
91. Patent №2175158 from 28 of April 2000
92. Patent №2219626 from 18 of July 2001
93. Patent №2239921 from 16 of June 2003
94. Patent №2399130 from 22 of January 2007
95. Patent №2400005 from 20 of May 2009

Index

V.V. Apollonov, *High-Power Optics*, Springer Series in Optical Sciences 192,
DOI 10.1007/978-3-319-10753-0

I

J

Q

MIX
Papier aus verantwortungsvollen Quellen
Paper from responsible sources
FSC® C105338

If you have any concerns about our products,
you can contact us on
ProductSafety@springernature.com

In case Publisher is established outside the EU,
the EU authorized representative is:
Springer Nature Customer Service Center GmbH
Europaplatz 3, 69115 Heidelberg, Germany

Printed by Libri Plureos GmbH
in Hamburg, Germany